Brainstorming Questions in Toxicology

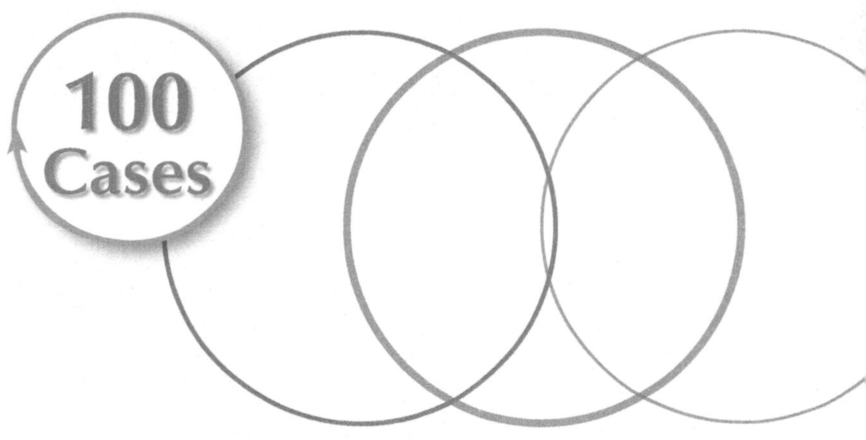

Brainstorming Questions in Toxicology

P. K. Gupta

CRC Press is an imprint of the
Taylor & Francis Group, an **informa** business

First edition published 2020
by CRC Press
6000 Broken Sound Parkway NW, Suite 300, Boca Raton, FL 33487-2742

and by CRC Press
2 Park Square, Milton Park, Abingdon, Oxon, OX14 4RN

CRC Press is an imprint of Taylor & Francis Group, LLC

ISBN: 978-0-367-89705-5 (hbk)
ISBN: 978-0-367-42952-2 (pbk)
ISBN: 978-1-003-00035-8 (ebk)

Typeset in Minion
by Lumina Datamatics Limited

DISCLAIMER

The information on questions and answers in this book is based on standard textbooks in the area of specialization. However, it is well-known that with the advancement of science the standard of care in the practice of toxicology changes rapidly. Though all the efforts have been made to ensure the accuracy of the information, the possibility of human error still remains. Therefore, neither the author nor the publisher guarantees that the information contained in the book is absolute. Anyone using the information contained in this book has to be, therefore, duly cautious. Neither the author nor the publisher should be responsible for any damage that results from the use of the information contained in any part of this book.

CONTENTS

Section 9: Special topics

PREFACE

This book entitled *Brainstorming Questions in Toxicology* serves as a comprehensive and quick reference for various examinations. This book contains multiple-choice questions, true or false/correct or incorrect statements, and match the correct statement, which are widely used in entrance tests, competitive examinations, and at some places in certifying examinations. It is therefore of utmost importance to induct items that not only tests the knowledge, understanding, and application ability of the student but also helps in learning objectives, methods of self-study, and self-assessment. Solving such brainstorming questions could encourage the crystallization and assimilation of the fundamental principles and applications in drug toxicology, clinical pharmacology, clinical toxicology, and medical toxicology. Thus, this book will be a useful tool and essential guide for forensic toxicologists, environmentalists, and veterinarians as well as those who want to prepare for licensure and certification exams, persons seeking continuing education, etc. It will also serve as a refresher for academicians and professionals in the field.

With these objectives, the book has been divided into several chapters that cover the general principles of toxicology, disposition and kinetics, mechanistic toxicology, target and non-target-directed toxicity, and the toxic effects of various xenobiotics, such as pesticides, metals and micronutrients, non-metals, solvents, gases and vapours, poisonous and venomous organisms, toxicity from human-use drugs, plant toxins, food and feed contaminants, radiation and radioactive materials, adverse effects of pollution and ecotoxicology, forensic and clinical toxicology, adverse effects of calories, toxic effects of nanomaterials, occupational exposure, and residual drug toxicity. Each chapter presents objective questions and answers; thus, it is a unique book in toxicology that reflects the breadth and multidisciplinary nature of toxicology with an objective approach to the subject.

In brief, this book is an essential guide and has been specifically targeted for a very specific audience of students, teachers, and established toxicologists.

Therefore, the author believes that this book will serve the students, academic institutions, and industry as follows:

- It is an excellent contribution for the students who need a study aid for toxicology but wants more than a textbook because they need a self-testing regime.
- It will be a useful tool for the teachers of toxicology who need inspiration when composing questions for their students.
- It will also help established toxicologists test their own knowledge of understanding the subject matter.
- It will be useful at universities and colleges and in industry for in-house training courses in toxicology, which I know exist in some pharmaceutical and chemical companies.
- It is required for all those who want to study for the toxicology boards and other examinations.

Thus, the main strength of this book is to improve the engagement and understanding of the subject.

Toxicology is a rapidly evolving field. Suggestions and comments are welcome to help the author improve the contents of the book. Please also suggest the deficiencies that need to be covered at drpkg_brly@yahoo.co.in or drpkg1943@gmail.com if you have any topics you feel should be better covered in any future editions.

P. K. Gupta

AUTHOR

P. K. Gupta is an internationally known toxicologist with more than 54 years of experience in teaching, research, and research management in the field of toxicology. To his credit, Dr. Gupta has several books and book chapters (John Wiley & Sons; Elsevier, Academic Press; Merck & Co., Mariel Limited; Springer Nature, to name a few) as well as scientific research publications (715) published in national and international peer-reviewed journals of repute. He has been a book review editor at Marcel Dekker, New York; an expert member consultant and advisor to WHO, Geneva; a consultant at the United Nations FAO, Rome, and International Atomic Energy Agency (IAEA), Vienna. He has been the founder and past president of the Society of Toxicology of India; founder, director, and member of the nominating committee of the International Union of Toxicology; and founder and editor-in-chief of the peer-reviewed PUBMED-indexed scientific journal *Toxicology International*. In addition, Dr. Gupta has also been the biographer for several editions of *WHO's WHO* from all over the world, including *Marquis WHO's WHO* (USA), IBC (UK), and other leading publications. His name appears in the *UP Book of Records* and *Limca Book of Records* for his unique scientific contribution in toxicology. At present, Dr. Gupta is the director of the Toxicology Consulting Group and president of the Academy of Sciences for Animal Welfare.

SECTION 1
GENERAL TOXICOLOGY

CHAPTER 1

PRINCIPLES OF TOXICOLOGY

1.1 MULTIPLE-CHOICE QUESTIONS

(Choose the most appropriate response.)

Exercise 1

Q.1. 'It is the dose that differentiates a toxicant from a poison.' This statement was made by scientist ________.
  a) Paracelsus
  b) Hippocrates
  c) Socrates
  d) Homer

Q.2. The branch of science that deals with assessing the toxicity of substances of plant and animal origin and those produced by pathogenic bacteria is ________.
  a) toxicology
  b) toxinology
  c) toxicokinetics
  d) toxicodynamics

Q.3. Minimum dose of a toxicant producing the desired response is called ________.
  a) ceiling dose
  b) threshold dose
  c) both a and b
  d) none

Q.4. The Arthus reaction is seen in the hypersensitivity of ________.
  a) type I
  b) type II
  c) type III
  d) type IV

Q.5. The Coolie breeds of dogs are hypersensitive to ________.
  a) Albendazole
  b) Ivermectin
  c) both
  d) none

Q.6. The measure of the margin of safety of a toxicant is obtained by ________.
  a) LD50/ED 99
  b) LD 1/ED 99
  c) ED50/LD 50
  d) LD50/ED 50

Q.7. A substance is called as moderately toxic if its median lethal dose is ________.
  a) 1–5 mg
  b) 5–500 mg
  c) 0.5–1 g
  d) >1 g

Q.8. A toxic substance produced by a biological system is specially referred to as a ________.
  a) toxicant
  b) toxin

c) xenobiotic
d) poison

Q.9. Allergic contact dermatitis is ________.
a) a non-immune response caused by a direct action of an agent on the skin
b) an immediate type-I hypersensitivity reaction
c) a delayed type-IV hypersensitivity reaction
d) characterized by the intensity of reaction being proportional to the elicitation dose
e) not involved in photoallergic reactions

Q.10. The reference dose (RfD) is generally determined by applying which of the following default procedures?
a) An uncertainty factor of 100 is applied to the NOAEL in chronic animal studies
b) A risk factor of 1000 is applied to the NOAEL in chronic animal studies
c) A risk factor of 10,000 is applied to the NOAEL in subchronic animal studies
d) An uncertainty factor between 10,000 and 1 million is applied to the NOAEL from chronic animal studies
e) Multiplying the NOAEL from chronic animal studies by 100

Q.11. Which of the following concerning the use of the 'benchmark dose' in risk assessment is *not* correct?
a) Can use the full range of doses and responses studied
b) Allows use of data obtained from experiments where a clear 'no observed adverse effect level' (NOAEL) has been attained
c) May be defined as the lower confidence limit on the 10% effective dose
d) Is primarily used for analyses of carcinogenicity data and has limited utility for analyses of developmental and reproduction studies that generate quantal data
e) Is not limited to the values of the administered doses

Q.12. Administration by oral gavage of a test compound that is highly metabolized by the liver versus subcutaneous injection will most likely result in ________.
a) less parent compound present in the systemic circulation
b) more local irritation at the site of administration caused by the compound
c) lower levels of metabolites in the systemic circulation
d) more systemic toxicity
e) less systemic toxicity

Q.13. The phrase that best defines 'toxicodynamics' is the ________.
a) linkage between exposure and dose
b) linkage between dose and response
c) dynamic nature of toxic effects among various species
d) dose range between desired biological effects and adverse health effects
e) loss of dynamic hearing range due to a toxic exposure

Q.14. Which of the following was banned under the Delaney Clause of the Food Additive Amendment of 1958?
a) Butylated hydroxytoluene
b) Sulfamethazine
c) Cyclamate
d) Phytoestrogens
e) Aflatoxin

Q.15. Which of the following toxicity can occur from a single exposure?
a) Acute toxicity
b) Subacute toxicity
c) Subchronic toxicity
d) Chronic toxicity

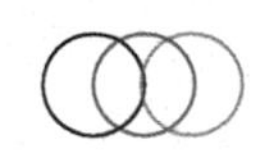

Q.16. Which of the following assumptions is *not* correct regarding risk assessment for male reproductive effects in the absence of mechanistic data?
a) An agent that produces an adverse reproductive effect in experimental animals is assumed to pose a potential reproductive hazard to humans.
b) In general, a non-threshold is assumed for the dose-response curve for male reproductive toxicity.
c) Effects of xenobiotics on male reproduction are assumed to be similar across species unless demonstrated otherwise.
d) The most sensitive species should be used to estimate human risk.
e) Reproductive processes are similar across mammalian species.

Q.17. A newly formed hapten protein complex usually stimulates the formation of a significant amount of antibodies in ________.
a) 1–2 min
b) 1–2 hours
c) 1–2 days
d) 1–2 weeks

Q.18. Prolonged muscle relaxation after succinylcholine is an example of a/an ________.
a) IGE-mediated allergic reaction
b) idiosyncratic reaction
c) immune complex reaction
d) reaction related to a genetic increase in the activity of a liver enzyme

Q.19. Increased production of methemoglobin is due to decreased activity of ________.
a) cytochrome P450 2B6
b) NADH cytochrome b5 reductase
c) cytochrome oxidase
d) cytochrome a3

Q.20. The most common target organ of toxicity is the ________.
a) heart
b) lung
c) CNS (brain and spinal cord)
d) skin

Q.21. The organs *least* involved in systemic toxicity are ________.
a) brain and peripheral nerves
b) muscle and bone
c) liver and kidney
d) hematopoietic system and lungs

Q.22. If two organophosphate insecticides are absorbed into an organism, the result will be ________.
a) additive effect
b) synergy effect
c) potentiation
d) subtraction effect

Q.23. If ethanol and carbon-tetrachloride are chronically absorbed into an organism, the effect on the liver would be ________.
a) additive effect
b) synergy
c) potentiation
d) subtraction effect

Q.24. If propyl alcohol and carbon tetrachloride are chronically absorbed into an organism, the effect on the liver would be ________.
a) additive effect
b) synergy

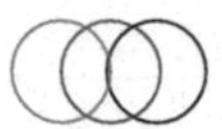

c) potentiation
d) subtraction effect

Q.25. The treatment of strychnine-induced convulsions by diazepam is an example of ________.
a) chemical antagonism
b) dispositional antagonism
c) receptor antagonism
d) functional antagonism

Q.26. The use of an antitoxin in the treatment of a snakebite is an example of ________.
a) dispositional antagonism
b) chemical antagonism
c) receptor antagonism
d) functional antagonism

Q.27. The use of charcoal to prevent the absorption of diazepam is an example of ________.
a) dispositional antagonism
b) chemical antagonism
c) receptor antagonism
d) functional antagonism

Q.28. The use tamoxifen in certain breast cancers is an example of ________.
a) dispositional antagonism
b) chemical antagonism
c) receptor antagonism
d) functional antagonism

Q.29. Chemicals known to produce dispositional tolerances are ________.
a) benzene and xylene
b) trichloroethylene and methylene chloride
c) paraquat and diquat
d) carbon tetrachloride and cadmium

Q.30. The most rapid exposure to a chemical would occur through which of the following routes:
a) oral
b) subcutaneous
c) inhalation
d) intramuscular

Q.31. A chemical that is toxic to the brain but which is detoxified in the liver would be expected to be ________.
a) more toxic orally than intramuscularly
b) more toxic rectally than intravenously
c) more toxic via inhalation than orally
d) more toxic on the skin than intravenously

Q.32. The LD_{50} is calculated from ________.
a) a quantal dose-response curve
b) a hormesis dose-response curve
c) a graded dose-response curve
d) a log-log dose-response curve

Q.33. A U-shaped graded toxicity dose-response curve is seen in humans with ________.
a) pesticides
b) sedatives
c) opiates
d) vitamins

Q.34. The TD_1/ED_{99} is called the ________.
a) margin of safety
b) therapeutic index

c) potency ratio
d) efficacy ratio

Q.35. All of the following are reasons for selective toxicity *except* ________.
a) transport differences between cell
b) biochemical differences between cell
c) cytology of male neurons versus female neurons
d) cytology of plant cells versus animal cells

Q.36. Regulatory toxicology aims at guarding the public from dangerous chemical exposures and depends primarily on which form of study:
a) observational human studies
b) controlled laboratory animal studies
c) controlled human studies
d) environmental studies

Q.37. Risk from a public health perspective is best described as
a) an undesirable end point is reached
b) a possibility of a bad outcome
c) the likelihood of an unwanted outcome combined with uncertainty of when it will occur
d) a bad outcome is assured and its mechanism is well understood

Q.38. Which of the following statements is regarding risk analysis?
a) It is a field of study that has been around for the last century.
b) It was developed by the pharmaceutical companies in response to concerns over new medications.
c) It is a relatively new field of study, spurred by new technologically based risks.
d) It was largely a private-sector venture.

Q.39. Which of the following are tools used in risk analysis?
a) Toxicology
b) Epidemiology
c) Clinical trials
d) All of the above

Q.40. Which of the following are common end points?
a) Death
b) No observable effect level
c) No observable adverse effect level
d) Lowest observable adverse effect level
e) All of the above

Q.41. The LD_{50} is best described as which of the following:
a) the dose at which 50% of all test animals die
b) the dose at which 50% of the animals demonstrate a response to the chemical
c) the dose at which all of the test animals die
d) the dose at which at least one of the test animal dies

Q.42. The effective dose is best described as which of the following:
a) the dose at which 50% of all test animals die
b) the dose at which some of the animals demonstrate a response to the chemical
c) the dose at which all of the animals demonstrate a response to the chemical
d) the dose at which 50% of all test animals demonstrate a response to the chemical

Q.43. Extrapolation is best described as which of the following:
a) using known information to reach a conclusion
b) using known information to infer something about the unknown
c) using speculative information to infer something about the known
d) a 'best guess' approach

Q.44. Which of the following assumptions is *not* correct regarding risk assessment for male reproductive effects in the absence of mechanistic data?
   a) An agent that produces an adverse reproductive effect in experimental animals is assumed to pose a potential reproductive hazard to humans.
   b) In general, a non-threshold is assumed for the dose-response curve for male reproductive toxicity.
   c) Effects of xenobiotics on male reproduction are assumed to be similar across species unless demonstrated otherwise.
   d) The most sensitive species should be used to estimate human risk.
   e) Reproductive processes are similar across mammalian species.

Q.45. The therapeutic index is usually defined as ________.
   a) TD_{50}/LD_{50}
   b) ED_{50}/LD_{50}
   c) LD_{50}/ED_{50}
   d) ED_{50}/TD_{50}

Q.46. The acetylator phenotype is ________.
   a) not found in dogs
   b) found exclusively in Asians
   c) responsible for the toxicity of amines
   d) an inherited trait affecting a particular metabolic reaction
   e) associated with the HLA type

Q.47. The phenomenon of enzyme induction involves
   a) an increase in the synthesis of the enzyme
   b) an increase in the activity of the enzyme
   c) an increase in liver weight
   d) a change in the substrate specificity of the enzyme
   e) an increase in bile flow

Q.48. Piperonyl butoxide and phenobarbitone
   a) are used in experimental studies to induce and inhibit the microsomal enzymes
   b) have no effect on drug metabolism
   c) uncouple electron transport in the mitochondria
   d) inhibit and induce the monooxygenase enzymes

Q.49. The differences between species in susceptibility to the toxicity of chemicals are usually the result of differences in metabolism.
   a) true
   b) false

Q.50. The toxic effects of a chemical may be influenced by which of the following:
   1) kidney function
   2) body weight
   3) rate of metabolism of the compound
   4) time of day the chemical is administered

   Select the correct statement.
   a) 1, 2, and 3
   b) 1 and 3
   c) 2 and 4
   d) 3 only 4
   e) all four

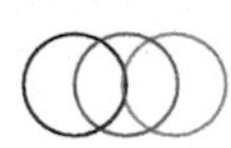

Answers (Exercise 1)

1. a	6. d	11. d	16. b	21. b	26. b	31. c	36. b	41. c	46. d
2. b	7. b	12. a	17. d	22. a	27. a	32. a	37. c	42. d	47. a
3. b	8. b	13. b	18. b	23. b	28. c	33. d	38. c	43. b	48. d
4. c	9. c	14. c	19. b	24. c	29. d	34. a	39. d	44. b	49. a
5. b	10. a	15. a	20. c	25. d	30. c	35. c	40. a	45. b	50. e

Exercise 2

Q.1. Examples of significant concentrations of a toxicant in a tissue that is not a target organ include all of the following *except* ________.
- a) lead in bone
- b) DDT in adipose tissue
- c) paraquat in lung
- d) TCDD in adipose tissue

Q.2. The ability of a chemical to cause acute skin and eye irritation is usually evaluated in a ________.
- a) rabbit
- b) rat
- c) mouse
- d) dog

Q.3. Before a potential pharmaceutical compound can be given to humans ________.
- a) an NDA must be filed with the FDA
- b) an IND must be filed with the FDA
- c) acute toxicity studies on four species must be conducted
- d) a two-year dog carcinogenicity study must be completed

Q.4. Phase 1 clinical trials are conducted to determine all of the following *except*________.
- a) pharmacokinetics
- b) safety
- c) rare adverse effects
- d) preliminary efficacy

Q.5. MTD stands for ________.
- a) minimum tolerated dose
- b) maximum total dose
- c) maximum tolerated dose
- d) maximum threshold dose

Q.6. The acute toxicity study in animals provides ________.
- a) an appropriate lethal dose
- b) information on target organs
- c) information on dose selection for long-term studies
- d) all of the above

Q.7. A subacute toxicity study in rats usually lasts ________.
- a) 3 days
- b) 14 days
- c) 3 months
- d) 6 months

Q.8. The period of organogenesis in rats is ________.
- a) day 3–10
- b) day 7–17
- c) day 12–25
- d) day 17–56

Q.9. A dose of investigational toxicant that suppresses body weight gain slightly in a 90-day animal study is defined by some regulatory agencies to be ________.
a) LOAEL
b) NOAEL
c) MTD
d) reference dose

Q.10. A subchronic animal study required by the FDA will usually include ________.
a) two species (usually one rodent and one non-rodent)
b) both genders
c) at least three doses (low, intermediate, and high)
d) all of the above

Q.11. A dose of a Compound A is toxic to animals in vivo. Another Chemical B is not toxic when given at doses several orders of magnitude higher. But when the two are given together, the toxic response is greater than that of the given dose of A alone. That is ________.
a) antagonism
b) synergism
c) additivity
d) potentiation
e) none of the above

Q.12. Which information may be gained from an acute toxicity study?
a) No effect level
b) LD50
c) Therapeutic index
d) Target organ
e) All of the above

Q.13. 1000 ppm is equivalent to 1%.
a) True
b) False

Q.14. Which one of the following statements is regarding toxicology?
a) Modern toxicology is concerned with the study of the adverse effects of chemicals on ancient forms of life.
b) Modern toxicology studies embrace principles from such disciplines as biochemistry, botany, chemistry, physiology, and physics.
c) Modern toxicology has its roots in the knowledge of plant and animal poisons, which predates recorded history and has been used to promote peace.
d) Modern toxicology studies the mechanisms by which inorganic chemicals produce advantageous as well as deleterious effects.
e) Modern toxicology is concerned with the study of chemicals in mammalian species.

Q.15. Knowledge of the toxicology of poisonous agents was published earliest in the ________.
a) *Ebers Papyrus*
b) *De Historia Plantarum*
c) *De Materia Medica*
d) *Lex Cornelia*
e) *Treatise on Poisons and Their Antidotes*

Q.16. Paracelsus, physician-alchemist, formulated many revolutionary reviews that remain integral to the structure of toxicology, pharmacology, and therapeutics today. He focused on the primary toxic agent as the chemical entity and articulated the dose-response relation. Which one of the following statements is not attributable to Paracelsus?
a) Natural poisons are quick in their onset of actions.
b) Experimentation is essential in the examination of responses to chemicals.
c) One should make the distinction between the therapeutic and toxic properties of chemicals.

 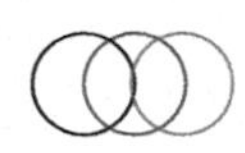

d) These properties are sometimes but not always indistinguishable except by dose.
e) One can ascertain a degree of specificity of chemicals and their therapeutic or toxic effects.

Q.17. The art of toxicology requires years of experience to acquire, even though the knowledge base facts may be learned more quickly. Which modern toxicologist is credited with saying that 'you can be toxicologist in two easy lessons, each of 10 years'?
a) Claude Bernard
b) Rachel Carson
c) Upton Sinclair
d) Arnold Lehman
e) Oswald Schmiedeberg

Q.18. Which of the following statements is correct?
a) Claude Bernard was a prolific scientist who trained over 120 students and published numerous contributions to the scientific literature.
b) Louis Lewin trained under Oswald Schmiedeberg and published much of the early work on the toxicity of narcotics, methanol, and chloroform.
c) An Introduction to the Study of Experimental Medicine was written by the Spanish physician Orfila.
d) Magendie used autopsy material and chemical analysis systematically as legal proof of poisoning.
e) Percival Potts was instrumental in demonstrating the chemical complexity of snake venoms.

Q.19. A concentration of 0.01% is equivalent to how many parts per million (ppm)?
a) 1 ppm
b) 10 ppm
c) 100 ppm
d) 1000 ppm
e) 10,000 ppm

Q.20. A blood lead concentration reported as 80 g/dL is the same as
a) 0.08 ppm
b) 0.8 ppm
c) 8 ppm
d) 80 ppm
e) 800 ppm

Q.21. If the toxic level of a toxicant in feed is 100 ppm for a 20-kg pig, what is the estimated toxicity of the toxicant on a milligram per kilogram of body weight basis? Assume the feed is air dried and the pig eats feed at the rate of 6% of its body weight daily.
a) 2 mg/kg
b) 4 mg/kg
c) 6 mg/kg
d) 8 mg/kg
e) 10 mg/kg

Q.22. Five identical experimental animals are treated with 1 mg of one of the following toxins. The animal treated with which toxin is most likely to die?
a) Ethyl alcohol (LD50 = 10,000 mg/kg)
b) Botulinum toxin (LD50 = 0.00001 mg/kg)
c) Nicotine (LD50 = 1 mg/kg
d) Ferrous sulfate (LD50 = 1500 mg/kg)
e) Picrotoxin (LD50 = 5 mg/kg)

Q.23. Place the following mechanisms of toxin delivery in order from most effective to least effective—1: intravenous; 2: subcutaneous; 3: oral; 4: inhalation; 5: dermal.
a) 1, 5, 2, 4, 3.
b) 4, 1, 2, 3, 5.

c) 1, 4, 2, 3, 5.
d) 4, 2, 1, 5, 3.
e) 1, 4, 3, 2, 5.

Q.24. A toxin with a half-life of 12 hours is administered every 12 hours. Which of the following is correct?
a) The chemical is eliminated from the body before the next dose is administered.
b) The concentration of the chemical in the body will slowly increase until the toxic concentration is attained.
c) A toxic level will not be reached, regardless of how many doses are administered.
d) Acute exposure to the chemical will produce immediate toxic effects.
e) The elimination rate of the toxin is much shorter than the dosing interval.

Q.25. Urushiol is the toxin found in poison ivy. It must first react and combine with proteins in the skin in order for the immune system to recognize and mount a response against it. Urushiol is an example of which of the following?
a) Antigen
b) Auto antibody
c) Superantigen
d) Hapten
e) Cytokine

Q.26. Toxic chemicals are most likely to be biotransformed in which of the following organs?
a) Central nervous system
b) Heart
c) Lung
d) Pancreas
e) Liver

Q.27. When Chemicals A and B are administered simultaneously, their combined effects are far greater than the sum of their effects when given alone. The chemical interaction between Chemicals A and B can be described as which of the following?
a) Potentiative
b) Additive
c) Antagonistic
d) Unconditionally antagonistic
e) Synergistic

Q.28. With respect to dose-response relationships, which of the following is correct?
a) Graded dose-response relationships are often referred to as 'all or nothing' responses.
b) Quantal dose-response relationships allow for the analysis of a population's response to varying dosages.
c) Quantal relationships characterize the response of an individual to varying dosages.
d) A quantal dose-response describes the response of an individual organism to varying doses of a chemical.
e) The dose-response always increases as the dosage is increased.

Q.29. When considering the dose-response relationship or an essential substance ________.
a) there are rarely negative effects of ingesting too much
b) the curve is the same for all people
c) adverse responses increase in severity with increasing or decreasing dosages outside of the homeostatic range
d) the relationship is linear
e) deficiency will never cause more harm than overingestion

Q.30. The therapeutic index of a toxicant ________.
a) is the amount of a toxicant needed to cure an illness
b) is lower in toxicants that are relatively safer
c) describes the potency of a chemical in eliciting a desired response

 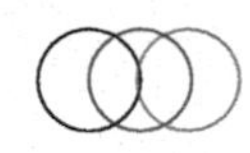

d) describes the ratio of the toxic dose to the therapeutic dose of a toxicant
e) explains the change in response to a toxicant as the dose is increased

Q.31. Penicillin interferes with the formation of peptidoglycan cross-links in bacterial cell walls, thus weakening the cell wall and eventually causing osmotic death of the bacterium. Which of the following is correct?
a) Treatment with penicillin is a good example of selective toxicity.
b) Penicillin interferes with human plasma membrane structure.
c) Penicillin is a good example of a toxicant with a low therapeutic index.
d) Penicillin is also effective in treating viral infections.
e) Penicillin is completely harmless to humans

Q.32. Which of the following is *not* important in hazard identification?
a) Structure–activity analysis
b) In-vitro tests
c) Animal bioassays
d) Susceptibility
e) Epidemiology

Q.33. The systematic scientific characterization of adverse health effects resulting from human exposure to hazardous agents is the definition of ________.
a) risk
b) hazard control
c) risk assessment
d) risk communication
e) risk estimate

Q.34. Which of the following is *not* an objective of risk management?
a) Setting target levels or risk
b) Balancing risks and benefits
c) Calculating lethal dosages
d) Setting priorities or manufacturers
e) Estimating residual risks

Q.35. Which of the following is *not* a feature in the design of standard cancer bioassays?
a) More than one species
b) Both sexes
c) Near lifetime exposure
d) Approximately 50 animals per dose group
e) Same dose level for all groups

Q.36. Which of the following types of epidemiologic study is always retrospective?
a) Cohort
b) Cross-sectional
c) Case-control
d) Longitudinal
e) Exploratory

Q.37. Which of the following is defined as the highest non-statistically significant dose tested?
a) ED50
b) ED100
c) NOAEL
d) ADI
e) COAEL

Q.38. Which of the following represents the dose below which no additional increase in response is observed?
a) ED_{10}
b) LD50
c) RfC

d) Threshold
e) Significance level

Q.39. Which of the following is *not* needed to calculate the reference dose using the BMD method?
a) MF
b) Percent benchmark response
c) NOAEL
d) UF
e) Benchmark dose

Q.40. Virtually safe doses are described at which confidence level?
a) 90%
b) 95%
c) 99%
d) 99.9%
e) 99.99%

Q.41. Which of the following is *not* a modifying actor that can influence the likelihood of disease?
a) Age
b) Dose
c) Nutritional status
d) Gender
e) Genetic susceptibility

Q.42. The probability of an adverse outcome is defined as ________.
a) hazard
b) exposure ratio
c) risk
d) susceptibility
e) epidemiology

Q.43. Which of following is *not* a modifying factor that can influence the likelihood of disease?
a) Age
b) Dose
c) Nutritional status
d) Gender
e) Genetic susceptibility

Q.44. What is the default assumption in human-health risk assessment when male reproductive effects are observed in animal studies and no mechanistic data are available?
a) An agent that produces an adverse reproductive effect in experimental animals is assumed to pose a potential reproductive hazard to humans.
b) A non-threshold is assumed for the dose-response curve for male reproductive toxicity.
c) The effects of xenobiotics on male reproduction are assumed to diverge radically across mammalian species.
d) Rodent reproductive processes are too dissimilar to those of humans to be used for human-health risk assessment.

Q.45. What is defined as 'a measurable biochemical, physiological, or other alteration within an organism that indicates health impairment or disease'?
a) Biomarker of exposure
b) Biomarker of susceptibility
c) Biomarker of effect
d) Biomarker of disease

 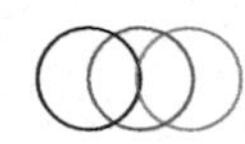

Q.46. In what way is a point of departure (POD) from an animal study usually modified in order to determine a reference dose (RfD)?

a) An uncertainty factor of 100 is applied to the NOAEL in chronic animal studies.
b) A risk factor of 1000 is applied to the NOAEL in chronic animal studies.
c) A risk factor of 10,000 is applied to the NOAEL in subchronic animal studies.
d) Multiplying the NOAEL from chronic animal studies by 100.

Q.47. What is a limitation of the benchmark dose approach in risk assessment?

a) Does not use the full range of doses and responses studied
b) Cannot be used when a clear 'no observed adverse effect level' (NOAEL) has already been attained
c) Is based on a predefined benchmark response that is arbitrary
d) Cannot be used to extrapolate beyond the range of administered doses

Q.48. What characteristic of perceived risk renders it difficult to quantify?

a) Rarely incorporated into risk communication decisions
b) Based on the precautionary principle
c) Highly influenced by factors such as familiarity and controllability
d) Too subjective

Answers (Exercise 2)

1. c	2. a	3. b	4. c	5. c	6. d	7. b	8. b	9. c	10. d	11. d	12. e
13. b	14. b	15. a	16. a	17. d	18. b	19. c	20. b	21. c	22. b	23. c	24. b
25. d	26. e	27. e	28. b	29. c	30. d	31. a	32. d	33. c	34. c	35. e	36. c
37. c	38. d	39. c	40. b	41. b	42. c	43. b	44. a	45. c	46. a	47. c	48. c

1.2 TRUE OR FALSE STATEMENTS

(Write T for True or F for False.)

Exercise 3

Q.1. Environmental risk is a well-understood entity.
Q.2. Cross-sectional studies look at the exposure and disease at the same time.
Q.3. Bias is a problem primarily of clinical trials.
Q.4. Subchronic studies are shorter than acute studies.
Q.5. The lethal dose refers to the dose at which 50% of test animals die.
Q.6. The maximum tolerated dose (MTD) is the level of chemical exposure where 10% of the animals die.
Q.7. Case control studies start with the exposure and follow for the disease.
Q.8. Case control studies are good for rare diseases.
Q.9. Clinical trials look at dose-response in animals.
Q.10. Dose refers to the amount of a substance in the environment.

Answers (Exercise 3)

1. F 2. T 3. F 4. F 5. T 6. F 7. F 8. T 9. F 10. F

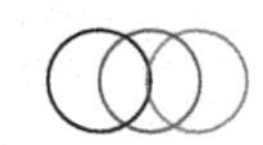

1.3 FILL IN THE BLANKS

Exercise 4

Q.1. The branch of science that deals with the harmful effects of physical and chemical agents of human and animal life is ________.

Q.2. In the term *toxicology*, the word *toxicon* (Greek) means ________.

Q.3. The branch of toxicology that deals with diagnosis, treatment, and management of toxic substances is known as ________.

Q.4. The development and interpretation of mandatory toxicology testing programs is addressed by ________.

Q.5. Investigating and controlling the toxic effects of various substances on the community is dealt by ________.

Q.6. The study of toxicity produced by substances of plant, animal, and microbial origin is termed as ________.

Q.7. A foreign chemical substance that is not normally produced in the body, or forms a part of the food, is known as ________.

Q.8. The source of adverse effect/damage is known as ________.

Q.9. The likelihood/probability of adverse effect upon exposure to a hazard is known as ________.

Q.10. The statement, 'all substances are poisons; the dose differentiates a poison from a remedy' is associated with ________.

Q.11. The scientist referred to as 'Father of Toxicology' is ________.

Q.12. DDT (dichloro-diphenyl-trichloro-ethane), which is used to control malaria and typhus, was discovered by ________.

Q.13. The person who is known as the 'Father of Nerve Agents' is ________.

Q.14. The author of the book *Silent Spring* in which the detrimental effect of DDT and other pesticides on the environment, particularly on birds, was documented is ________.

Q.15. The Bhopal gas tragedy, which is considered to be the world's worst industrial disaster, was caused due to the leakage of ________ from Union Carbide fertilizer company.

Q.16. Use of thalidomide in pregnant women for treating morning sickness led to ________ condition in the infants.

Q.17. If the action of one substance opposes or neutralizes the effect of another substance, the relationship is referred to as ________.

Q.18. Rodents are preferred for oral toxicity testing because they lack ________ reflex.

Q.19. The maximum acceptable/permitted amount of a drug present in feed and foods is known as ________.

Q.20. The highest dose of a compound that produces adverse effects but no mortality is called ________.

Q.21. If the period of exposure of a toxicant is more than 3 months, the type of study is termed as ________.

Q.22. The type of toxicity that results due to the progressive accumulation of a toxicant in the body is known as ________.

Q.23. The amount of toxicant in food and water that can be consumed daily over a lifetime without any significant health risk is called ________.

Q.24. The unlawful or criminal killing of animals through the administration of poisons is known as ________.

Q.25. The unintentional addition of toxicants and contaminants in feed and water is known as ________.

Q.26. Man-made sources of toxicants are referred to as ________ sources.

Q.27. Failure to elicit a response to an ordinary dose of a substance due to prior usage is known as ________.

Q.28. The beneficial effects of toxic substances at low doses are known as ________.

Q.29. In the event of irreparable injury, the cell undergoes a process of programmed cell death known as ________.

Q.30. A substance is classified as extremely toxic if the lethal dose (LD) is less than ________ and as practically non-toxic if the LD is ________.

Answers (Exercise 4)

1. Toxicology
2. Poison
3. Clinical toxicology
4. Regulatory toxicology
5. Toxicovigilance
6. Toxinology
7. Xenbiotic
8. Hazard
9. Risk
10. Paracelsus
11. M.J.B. Orfila
12. Pau Muller
13. Gerhard Schrader
14. Rachel Carson
15. Methyl-isocyanate (MIC)
16. Phocomelia
17. Antagonism
18. Vomition
19. Maximum residue level (MRL)
20. Maximum tolerated dose or LD_0
21. Chronic toxicity
22. Cumulative toxicity
23. Acceptable daily intake (ADI)
24. Malicious poisoning
25. Accidental poisoning
26. Anthropogenic
27. Tolerance
28. Hormesis
29. Apoptosis
30. 1 mg/kg, 5–15 g/kg

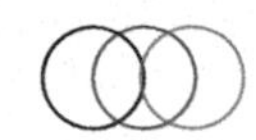

1.4 MATCH THE STATEMENTS

(Column A with Column B)

Exercise 5

Column A		Column B	
Q.1.	Hippocrates	a.	corrosive
Q.2.	MJB Orfila	b.	less than 5 mg/kg
Q.3.	nutritional toxicology	c.	age
Q.4.	genetic	d.	father of medicine
Q.5.	sulfuric acid	e.	father of toxicology
Q.6.	super toxic	h.	food/feed stuffs
Q.7.	infants	i.	hereditary
Q.8.	tolerance	j.	safety
Q.9.	therapeutic index	k.	receptor
Q.10.	interaction	l.	toxic reaction

Answers (Exercise 5)

1. d 2. e 3. h 4. i 5. a 6. b 7. c 8. l 9. j 10. k

Exercise 6

Column A		Column B	
Q.1.	antagonism	a.	program to prevent birth defects
Q.2.	TOCP	b.	programs human risk assessment
Q.3.	probit unit	c.	idiosyncratic-prolonged apnoea
Q.4.	synergy	d.	delayed neurotoxicity
Q.5.	succinylcholine	e.	normal equivalent derivation plus 5
Q.6.	STEPS	f.	$4+0=1$
Q.7.	Superfund Act	g.	toxicology and the law
Q.8.	descriptive toxicology	h.	$2+3=10$
Q.9.	regulatory toxicology	i.	performs toxicology testing
Q.10.	forensic toxicology	j.	studies/treats human disease caused by toxins
Q.11.	clinical toxicology	k.	support to clean toxic-waste sites
Q.12.	organophosphate	l.	ARfD
Q.13.	methyl eugenol	m.	MOE
Q.14.	methylmercury	n.	TDI
Q.15.	E300 (ascorbic acid)	o.	ADI

Answers (Exercise 6)

1. f 2. d 3. e 4. h 5. c 6. a 7. k 8. i 9. b 10. g
11. j 12. o 13. n 14. m 15. l

Chapter 2

DISPOSITION AND TOXICOKINETICS

2.1 MULTIPLE-CHOICE QUESTIONS

(Choose the most appropriate response.)

Exercise 1

Q.1. Biotransformation is vital in removing toxins from the circulation. All of the following statements regarding biotransformation are correct *except*

a) Many toxins must be biotransformed into a more lipid-soluble form before they can be excreted from the body.
b) The liver is the most active organ in the biotransformation of toxins.
c) Water solubility is required in order for many toxins to be excreted by the kidney.
d) The kidney plays a major role in eliminating toxicants from the body.
e) The lungs play a minor role in ridding the body of certain types of toxins.

Q.2. Which of the following statements about active transport across cell membranes is *false*?

a) Unlike simple or facilitated diffusion, active transport pumps chemicals against an electrochemical or concentration gradient.
b) Unlike simple diffusion, there is a rate at which active transport becomes saturated and cannot move chemicals any faster.
c) Active transport requires the expenditure of ATP in order to move chemicals against electrochemical or concentration gradients.
d) Active transport exhibits a high level of specificity of the compounds that are being moved.
e) Metabolic inhibitors do not affect the ability to perform active transport.

Q.3. Which of the following might increase the toxicity of a toxin administered orally?

a) Increased activity of the MDR transporter (p-glycoprotein)
b) Increased biotransformation of the toxin by gastrointestinal cells
c) Increased excretion of the toxin by the liver into bile
d) Increased dilution of the toxin dose
e) Increased intestinal motility

Q.4. Which of the following most correctly describes the first-pass effect?

a) The body is most sensitive to a toxin the first time that it passes through the circulation.
b) Orally administered toxins are partially removed by the GI tract before they reach the systemic circulation.
c) It only results from increased absorption of toxins by GI cells.
d) It is often referred to as 'postsystemic elimination'.
e) A majority of the toxin is excreted after the first time the blood is filtered by the kidneys.

Q.5. Which of the following is an important mechanism of removing particulate matter from the alveoli?

a) Coughing
b) Sneezing
c) Blowing one's nose
d) Absorption into the bloodstream, followed by excretion via the kidneys
e) Swallowing

Q.6. For a toxin to be absorbed through the skin, it must pass through multiple layers in order to reach the systemic circulation. Which of the following layers is the most important in slowing the rate of toxin absorption through the skin?
a) Stratum granulosum
b) Stratum spinosum
c) Stratum corneum
d) Stratum basale
e) Dermis

Q.7. A toxin is selectively toxic to the lungs. Which of the following modes of toxin delivery would most likely cause the *least* damage to the lungs?
a) Intravenous
b) Intramuscular
c) Intraperitoneal
d) Subcutaneous
e) Inhalation

Q.8. Which of the following is *not* an important site of toxicant storage in the body?
a) Adipose tissue
b) Bone
c) Plasma proteins
d) Muscle
e) Liver

Q.9. Which of the following regarding the blood–brain barrier is ________.
a) The brains of adults and newborns are equally susceptible to harmful blood-borne chemicals.
b) The degree of lipid solubility is a primary determinant in whether or not a substance can cross the blood–brain barrier.
c) Astrocytes play a role in increasing the permeability of the blood–brain barrier.
d) Active-transport processes increase the concentration of xenobiotics in the brain.
e) The capillary endothelial cells of the CNS possess large fenestrations in their basement membranes.

Q.10. Which of the following will result in DECREASED excretion of toxic compounds by the kidneys?
a) A toxic compound with a molecular weight of 25,000 Da
b) Increased activity of the multidrug-resistance (MDR) protein
c) Increased activity of the multiresistant drug protein (MRP)
d) Increased activity of the organic cation transporter
e) Increased hydrophilicity of the toxic compound

Q.11. Biodistribution of nanoparticles may be influenced by ________.
a) physicochemical properties such as plasma protein and respiratory tract mucus
b) physicochemical properties such as surface size and chemistry
c) physicochemical properties such as the gastrointestinal milieu
d) body compartment media including surface hydrophobicity
e) body compartment media including size

Q.12. Toxicants most likely to be reabsorbed after being filtered at the glomerulus are ________.
a) organic anions
b) organic cations
c) natural polar molecules
d) highly lipid-soluble molecules

Q.13. A high urinary pH would favour the excretion of ________.
a) organic acids
b) organic bases

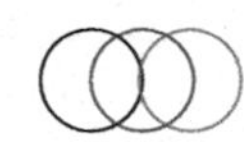

c) neutral organic compounds
d) none of the above

Q.14. Diuretics can enhance the renal elimination of compounds that ________.
a) are of molecular weight greater than 70 kDa
b) are ions trapped in the tubular lumen
c) are highly lipid soluble
d) are highly protein bound

Q.15. The amount of a volatile liquid excreted by the lungs is ________.
a) inversely proportional to its lipid-water partition coefficient
b) directly proportional to its pressure
c) directly proportional to its molecular weight
d) inversely proportionate to cardiac output

Q.16. Kernicterus results from ________.
a) enzyme induction leading to decreased glucocorticoid levels
b) excess ingestion of foods containing tyramine
c) displacement of bilirubin from plasma proteins
d) malabsorption of fat-soluble vitamins

Q.17. All of the following could influence the gastrointestinal absorption of xenobiotics *except* ________.
a) pH
b) intestinal microflora
c) presence of food
d) time of day

Q.18. The rate of diffusion of a xenobiotic across the GI tract is proportional to all of the following *except* ________.
a) hepatic blood flow
b) surface area
c) permeability
d) residence time

Q.19. Which of the following is *not* absorbed in the colon?
a) Water
b) Sodium ion
c) Glucose
d) Hydrogen ion

Q.20. Nanoparticles are considered to have diameter smaller than ________.
a) 100 μm
b) 10 μm
c) 1 μm
d) 0.1 μm

Q.21. All of the following are true of nanoparticles *except* ________.
a) They are capable of exposing the lung to a large number of particles.
b) They are capable of exposing the lung to a large particle surface area.
c) Because of turbulence, very few reach the alveoli.
d) They are the focus of recent toxicologic research.

Q.22. The first-pass effect for most of the drugs occurs in the ________.
a) tongue
b) intestines
c) rectum
d) oral mucosa

Q.23. An example for "lethal synthesis" is the conversion of ________.
a) codeine to morphine
b) parathion to paraoxon

c) phenylbutazone to oxyphenbutazone
d) vitamin K to vitamin K epoxide

Q.24. The area under the curve (AUC) denotes the value of ________.
a) volume of distribution of the drug
b) bioavailability of the drug
c) half-life of the drug
d) maximum plasma concentration of the drug

Q.25. An antagonist of warfarin is ________.
a) protamine sulfate
b) clopidogrel
c) phytomenadione
d) ethamsylate

Q.26. An example for synergistic antibacterial combination is ________.
a) chloramphenicol + ampicillin
b) tetracycline + fluororquinolone
c) tylosin + lincomycin
d) cefazolin + gentamicin

Q.27. The major site of absorption of poisonous substances for monogastric animals is ________.
a) large intestine
b) small intestine
c) colon
d) rectum

Q.28. Movement of xenobiotics molecules from the point of exposure site into circulation is called ________.
a) absorption
b) distribution
c) metabolism
d) excretion

Q.29. Acetylation conjugation is absent in ________.
a) cats
b) dogs
c) pigs
d) horses

Q.30. Glucuronide conjugation is absent in ________.
a) cats
b) dogs
c) pigs
d) horses

Q.31. Ethereal sulfate formation is absent in ________.
a) cats
b) dogs
c) pigs
d) horses

Q.32. All of the following can cross the placenta *except* ________.
a) heparin
b) rubella virus
c) spirochetes
d) IGG antibody

Q.33. Methylmercury crosses the blood–brain barrier by combining with cysteine and forming a molecular similar to ________.
a) glycine
b) glutamine

c) taurine
d) methionine

Q.34. Which of the following statements are correct?
a) The blood–brain barrier of a 70-year-old is more permeable than that of a premature infant.
b) Chemicals/drugs can be excreted into the urine by active secretion.
c) The kidney lacks cytochrome P450 enzymes.
d) All mammalian placentas have the same number of tissue layers.

Q.35. All of the following are correct of breast milk *except* ________.
a) acidic compounds may be more concentrated in milk than plasma.
b) toxicants can be passed from mother to offspring.
c) toxicants can be passed from cows to humans.
d) DDT, PCBs, and PBBs can be found in human milk.

Q.36. Active transport is characterized by all of the following *except* ________.
a) a movement against a concentration gradient
b) energy requirement
c) non-saturability
d) competitive inhibition

Q.37. All of the following are correct of the facilitated diffusion *except* ________.
a) does not require energy
b) movement against a concentration gradient
c) saturability
d) involvement of a carrier

Q.38. Which of the following does *not* uncouple oxidative phosphorylation?
a) Pentachlorophenol
b) Dinitrophenol
c) Aconitase
d) Salicylate
e) Gramicidin

Q.39. Which of the following is *not* regarding peroxisome proliferators?
a) They are a structurally diverse group of chemicals.
b) They cause marked induction of lipid-metabolizing enzymes.
c) They often are non-genotoxic hepatocarcinogens in rodents.
d) They induce hepatic CYP1A, which is indicative of peroxisome proliferation.
e) They operate via the peroxisome-proliferator-activated receptor.

Q.40. For which of the following routes of exposure is pre-systemic elimination possible?
a) Oral (GIT)
b) Inhalation
c) Intramuscular
d) Intravenous

Q.41. All of the following are significantly stored in the bone matrix *except* ________.
a) lead
b) diquat
c) strontium
d) fluoride

Q.42. Of the more than 40 cytochrome P450 isozymes, which six account for the majority of xenobiotic metabolism in humans?
a) CYP1E1, CYP2B1, CYP2B19, CYP2F1, CYP3A7, CYP4A6
b) CYP1A1, CYP2A6, CYP2B6, CYP2D6, CYP2F2, CYP4B2
c) CYP1B2, CYP2C6, CYP2F2, CYP3A2, CYP3A4, CYP4A2
d) CYP1A2, CYP2C9, CYP2C19, CYP2D6, CYP2E1, CYP3A4
e) CYP1A1, CYP2C8, CYP2F2, CYP3A7, CYP4A2, CYP4A6

Q.43. Which of the following is *not* a characteristic of active transport?
a) Blocked by saxitoxin
b) Movement against a concentration gradient
c) Exhibits a transport maximum
d) Energy dependent
e) Selectivity

Q.44. Which of the following does *not* inhibit electron transport?
a) Rotenone
b) Succinate
c) Antimycin-A
d) Formate
e) Azide

Q.45. Reabsorption of toxicants does *not* occur through ________.
a) entero-hepatic recirculation
b) glomerular filtration
c) diffusion
d) active transport
e) carriers for physiologic oxyanions

Q.46. When all receptors are occupied by a toxicant and there is a maximum amount of receptor–toxicant complexes, the response is labelled ________.
a) $t_{1/2}$
b) LC_{max}
c) E_{max}
d) C_{max}

Q.47. Which of the following statements is correct?
a) Toxicant receptor interactions are always reversible.
b) Receptors for toxicants are always enzymes.
c) The toxic response is related to the toxicant concentration in the plasma more so than the concentration at the site of action.
d) none of the above

Q.48. An increase in free drug concentration will ________.
a) increase the pharmacological effect
b) decrease the toxic effect
c) decrease the amount of drug filtered at the glomerulus
d) none of the above

Q.49. The phrase that best defines "toxicodynamics" is the ________.
a) linkage between exposure and dose
b) linkage between dose and response
c) dynamic nature of toxic effects among various species
d) dose range between desired biological effects and adverse health effects
e) loss of dynamic hearing range due to a toxic exposure

Q.50. Methylation ________.
a) typically increases the water solubility of xenobiotics
b) is a major pathway of xenobiotic metabolism
c) requires S-adenosylmethionine (SAM)
d) requires acetyl coenzyme A
e) requires taurine

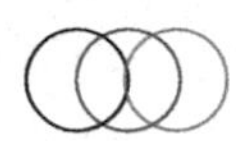

Answers (Exercise 1)

1. a	6. c	11. b	16. c	21. c	26. d	31. a	36. c	41. b	46. c
2. e	7. c	12. d	17. d	22. b	27. a	32. a	37. b	42. d	47. d
3. d	8. d	13. a	18. a	23. b	28. b	33. d	38. c	43. a	48. a
4. b	9. b	14. b	19. c	24. b	29. a	34. b	39. d	44. b	49. b
5. e	10. d	15. b	20. d	25. c	30. b	35. a	40. a	45. b	50. c

Exercise 2

Q.1. A probe drug for human CYP2C19 activity is ________.
- a) mephenytoin
- b) valproic acid
- c) carbamazepine
- d) warfarin

Q.2. All of the following are true of CYP2D6 *except* ________.
- a) It converts codeine to morphine.
- b) It is polymorphic.
- c) It is induced by quinidine.
- d) Poor metabolizers have a lower risk of lung cancer.

Q.3. The aryl hydrocarbon receptor agonist includes all of the following *except* ________.
- a) TCDD
- b) benzopyrene
- c) 3-methylcholanthrene
- d) benzene

Q.4. Enzyme induction in humans has been associated with ________.
- a) osteomalacia
- b) hepatocellular carcinoma
- c) cirrhosis
- d) psoriasis

Q.5. In the metabolism-dependent inhibition of cytochrome P450 ________.
- a) the parent compound is a potent inhibitor
- b) the metabolite must be a product of P450 catalysis
- c) the metabolite is a potent inhibitor
- d) the inhibition is always irreversible

Q.6. A compound that induces CYP2D6 is ________.
- a) rifampin
- b) dexamethazone
- c) ethanol
- d) none of the above

Q.7. All of the following are considered Phase I biotransformation reactions *except* ________.
- a) hydrolysis
- b) conjugation
- c) reduction
- d) oxidation

Q.8. All of the following statements are correct *except* ________.
- a) Forms of epoxide hydrolase can exist in both microsomes and cytosol.
- b) Gemfibrozil is conjugated with glucuronic acid before it is oxidized by cytochrome P 50.
- c) CYP2D6 and CYP2C9 metabolize over half of the drugs in current use.
- d) Biotransformation can take place in the gut.

Q.9. UDP glucuronyltransferases conjugate all of the following endogenous molecules *except* ________.
a) thyroid hormone
b) steroid hormones
c) parathyroid hormone

Q.10. If codeine were given to a patient who was a CYP2D6 poor metabolizer, the most likely result would be ________.
a) inadequate analgesia
b) higher-than-normal levels of morphine at 2 hours post-dose
c) higher-than-normal levels of codeine at 4 hours post-dose
d) higher-than-normal levels of oxycodone at 4 hours post-dose

Q.11. A victim drug is ________.
a) a drug whose clearance is determined mostly by a single route of administration
b) a drug that induces neutralizing antibodies
c) a drug that is unstable in plasma
d) a racemic drug mixture where one isomer inhibits the metabolism of the other isomer

Q.12. Terfenadine and ketoconazole are examples of ________.
a) enzyme inducers
b) perpetrator and inhibitor
c) victim drug and perpetrator
d) drugs with limited biotransformation

Q.13. All of the following are correct *except* ________.
a) hyperforin induces CYP3A4
b) broccoli inhibits 1A2
c) grapefruit juice inhibits intestinal CYP3A4
d) drugs that inhibit transporters can help anticancer agents

Q.14. Levels of UDPGA and PAPS are lowered by ________.
a) St. John's Wort
b) phenobarbital
c) rifampin
d) fasting

Q.15. An example of a pair of enantiomers in which one inhibits CYP2D6 and the other has little inhibiting activity is ________.
a) R and S methadone
b) R and S warfarin
c) R and S mephenytoin
d) quinidine and quinine

Q.16. Which of the following biotransformation enzyme-subcellular location pairs is correct?
a) Alkaline phosphatase-cell membrane
b) Carboxylesterase-blood
c) Sulfotransferase-cytosol
d) All of the above

Q.17. The least likely biotransformation reaction that aniline would undergo is ________.
a) halogenation
b) aromatic hydroxylation
c) *N*-acetylation
d) *N*-glucuronidation

Q.18. The proteins KEAPI and Nrf2 ________.
a) suppress CYP expression in response to inflammation
b) induce enzymes in response to oxidative stress
c) promote DNA methylations
d) none of the above

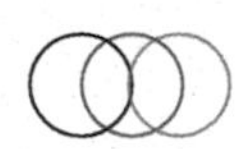

Q.19. All of the following are correct about glutathione *except* ________.
a) germ cells and ovum have high levels
b) conjugation of dibromoethane results in a mutagenic metabolite
c) conjugation of electrophiles is a major means of protecting DNA
d) conjugation always occurs enzymatically

Q.20. Phenobarbital ________.
a) causes liver tumours in humans and rodents
b) causes liver tumours in rodents but not humans
c) causes liver tumours in primates but not in rodents
d) causes liver tumours in rodents and nasal tumours in humans

Q.21. The systemic availability of an orally administered toxicant is dependent on ________.
a) gastrointestinal absorption
b) intestinal mucosa metabolism
c) first-pass liver metabolism
d) all of the above

Q.22. Which of the following statements are correct?
a) Renal clearance is equal to urine formation.
b) Hepatic clearance cannot exceed hepatic blood flow.
c) A process that increases free drug concentration will decrease hepatic clearance and increase renal clearance.
d) Total body clearance equals a dose divided by the half-life.

Q.23. A classic example of a drug inducing its own metabolism is ________.
a) warfarin
b) lovastatin
c) carbamazepine
d) theophylline

Q.24. An example of a thermodynamic parameter used in physiologic, toxicokinetic models is ________.
a) tissue partition coefficient
b) alveolar ventricular rate
c) cardiac output
d) liver volume

Q.25. Fick's Law of Diffusion ________.
a) is a zero-order process
b) is a first-order process
c) applies to active transport
d) requires energy

Q.26. The method of predicting the toxicokinetic behaviour of chemicals and drugs across species is called ________.
a) Monte Carlo stimulation
b) benchmark kinetics
c) allometric scaling
d) linear regression kinetics

Q.27. Which of the following is *not* theoretically possible?
a) Volume of distribution greater than volume of human body
b) Volume of distribution equal to blood volume
c) Total clearance equal to renal clearance
d) Bioavailability (F) greater than 1

Q.28. A compartment in which the uptake of a xenobiotic is dependent on membrane permeability and total membrane area is called ________.
a) perfusion limited
b) diffusion limited

c) blood-flow limited
d) ventilation limited

Q.29. The alpha phase of an intravenously administered drug classically represents the ________.
a) absorption phase
b) elimination phase
c) dissolution phase
d) distribution phase

Q.30. An advantage of a physiologic, toxicokinetic model over a classic model is ________.
a) it may be able to predict tissue concentration
b) it only has two compartments
c) the mathematics are less complicated
d) it can give a better estimation of bioavailability

Answers (Exercise 2)

1. a	4. a	7. b	16. d	19. d	22. b	10. b	13. b	25. b	28. b
2. c	5. c	8. c	17. a	20. b	23. c	11. a	14. d	26. c	29. d
3. d	6. d	9. c	18. b	21. d	24. a	12. c	15. d	27. d	30. a

Exercise 3

Q.1. Xenobiotic biotransformation is performed by multiple enzymes in multiple subcellular locations. Where would one of these enzymes most likely *not* be located?
a) Cytosol
b) Golgi apparatus
c) Lysosome
d) Mitochondria
e) Microsome

Q.2. All of the following statements regarding hydrolysis, reduction, and oxidation biotransformations are true *except* ________.
a) The xenobiotic can be hydrolyzed.
b) The xenobiotic can be reduced.
c) There is a large increase in hydrophilicity.
d) The reactions introduce a functional group to the molecule.
e) The xenobiotic can be oxidized.

Q.3. Which of the following is often conjugated to xenobiotics during Phase II biotransformations?
a) Alcohol group
b) Sulfhydryl group
c) Sulfate group
d) Aldehyde group
e) Carbonyl group

Q.4. Which of the following is a statement about the biotransformation of ethanol?
a) Alcohol dehydrogenase is only present in the liver.
b) Ethanol is reduced to acetaldehyde by alcohol dehydrogenase.
c) Ethanol and hydrogen peroxide combine to form acetaldehyde with the aid of a catalase.
d) In spite of its catalytic versatility, cytochrome P450 does not aid in ethanol oxidation.
e) Acetaldehyde is oxidized to acetic acid in the mitochondria by aldehyde dehydrogenase.

Q.5. Which of the following enzymes is responsible for the biotransformation and elimination of serotonin?
a) Cytochrome P450
b) Monoamine oxidase

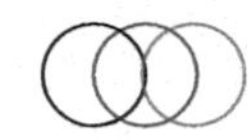

   c) Flavin monooxygenase
   d) Xanthine oxidase
   e) Paraoxonase

Q.6. Which of the following reactions would likely *not* be catalyzed by cytochrome P450?
   a) Dehydrogenation
   b) Oxidative group transfer
   c) Epoxidation
   d) Reductive dehalogenation
   e) Ester cleavage

Q.7. All of the following statements regarding cytochrome P450 are true *except* ________.
   a) Poor metabolism or biotransformation of xenobiotics is often due to a genetic deficiency in cytochrome P450.
   b) Cytochrome P450 can be inhibited by both competitive and non-competitive inhibitors.
   c) Certain cytochrome P450 enzymes can be induced by one's diet.
   d) Increased activity of cytochrome P450 always shows the rate of xenobiotic activation.
   e) Induction of cytochrome P450 can lead to increased drug tolerance.

Q.8. Which of the following statements regarding Phase II biotransformation (conjugation) reactions is ________?
   a) Phase II reactions greatly increase the hydrophilicity of the xenobiotic.
   b) Phase II reactions are usually the rate-determining step in the biotransformation and excretion of xenobiotics.
   c) Carboxyl groups are very common additions of Phase II reactions.
   d) Most Phase II reactions occur spontaneously.
   e) Increased Phase II reactions result in increased xenobiotic storage in adipocytes.

Q.9. Where do most Phase II biotransformations take place?
   a) Mitochondria
   b) ER
   c) Blood
   d) Nucleus
   e) Cytoplasm

Q.10. Which of the following is *not* an important co-substrate for Phase II biotransformation reactions?
   a) UDP-glucuronic acid
   b) 3′-phosphoadenosine-5′-phosphosulphate (PAPS)
   c) *S*-adenosyl methionine (SAM)
   d) *N*-nitrosodiethyl amine
   e) Acetyl CoA

Q.11. All of the following are hydrolytic enzymes *except* ________.
   a) carboxylesterase
   b) alcohol dehydrogenase
   c) cholinesterase
   d) paraoxonase

Q.12. Nitro reductase plays an important role in ________.
   a) nasal epithelium
   b) lung Clara cells
   c) white blood cells
   d) intestinal flora

Q.13. A drug that undergoes sulfoxide reduction is ________.
   a) haloperidol
   b) chloramphenicol
   c) thalidomide
   d) sulindac

Q.14. Quinidine oxidoreductases are thought to play a protective role in ________.
a) liver toxicity of microcystin
b) bone marrow toxicity of benzene
c) renal toxicity of aminoglycosides
d) neurotoxicity of *n*-hexane

Q.15. All of the following are mechanisms for removing halogen atoms aliphatic xenobiotics *except* ________.
a) Grignard dehalogenation
b) reductive dehalogenation
c) oxidative dehalogenation
d) double dehalogenation

Q.16. Oxidation of ethanol to acetaldehyde takes place in ________.
a) cytosol
b) microsomes
c) peroxisomes
d) all of the above

Q.17. Reductive dehalogenation of carbon tetrachloride produces ________.
a) phosgene
b) chloroform
c) trichloromethyl radical
d) hydrochloric acid

Q.18. Acetaldehyde is converted to acetic acid by ALDH2 in ________.
a) mitochondria
b) cytosol
c) microsomes
d) all of the above

Q.19. Aldehyde oxidase and xanthene oxidoreductase contain ________.
a) zinc
b) molybdenum
c) selenium
d) copper

Q.20. Slow acetylators of NAT demonstrate all of the following *except* ________.
a) peripheral neuropathy from isoniazid
b) systemic lupus erythematous from procainamide
c) peripheral neuropathy from dapsone
d) decreased hypotensive response from hydrazine

Q.21. In contrast to glucuronidation, sulfonation is ________.
a) a low-affinity, low-capacity pathway
b) a low-affinity, high-capacity pathway
c) a high-affinity, high-capacity pathway
d) a high-affinity, low-capacity pathway

Q.22. Induction of sulfotransferase enzymes by rifampin are clinically relevant for ________.
a) warfarin
b) digoxin
c) ethinyl estradiol
d) all of the above

Q.23. All of the following statements regarding sulfonation reactions are true *except* ________.
a) They can make a molecule less lipid soluble.
b) They always detoxify a molecule.
c) Some drugs must be metabolized to a sulfonate conjugate to have a pharmacologic effect.
d) Morphine-6-sulfate is more potent than morphine in the rat.

 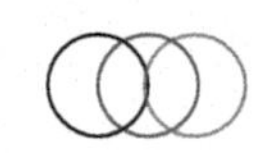

Q.24. All of the following statements are true regarding methylation *except* ________.
a) The process generally decreases the water solubility of the parent.
b) The process can mask functional groups that can be metabolized by other conjugation enzymes.
c) Inorganic mercury and arsenic can be dimethylated.
d) High methyl transferase activity may lower levels of homocysteine.

Q.25. All of the following are methyl transferase enzymes *except* ________.
a) SAM
b) COMT
c) NNMT
d) HNMT

Q.26. All of the following are of glucuronide conjugates of xenobiotics *except* ________.
a) They can be excreted in the urine.
b) They are formed from activated xenobiotics.
c) They are substances for beta-glucuronidase in the intestinal flora.
d) They can be excreted into bile.

Q.27. All of the following are true of the sulfonation reaction *except* ________.
a) They involve the transfer of sulfate.
b) They are catalyzed by sulfotransferases.
c) The cofactor of the reaction is PAPS.
d) They are mainly excreted in the urine.

Q.28. The number of UGT mammalian enzymes that have been identified is approximately ________.
a) 5
b) 12
c) 22
d) 58

Q.29. In addition to the cytoplasm, sulfotransferases are present in mammals in the ________.
a) endoplasmic reticulum
b) mitochondria
c) plasma membrane
d) Golgi apparatus

Q.30. Regarding the two-compartment model of classic toxicokinetics, which of the following is correct?
a) There is rapid equilibration of the chemical between central and peripheral compartments.
b) The logarithm of plasma concentration versus time data yields a linear relationship.
c) There is more than one dispositional phase.
d) It is assumed that the concentration of a chemical is the same throughout the body.
e) It is ineffective in determining effective doses in toxicity studies.

Q.31. When calculating the fraction of a dose remaining in the body over time, which of the following factors need *not* be taken into consideration?
a) Half-life
b) Initial concentration
c) Time
d) Present concentration
e) Elimination rate constant

Q.32. All of the following statements regarding the apparent volume of distribution (V_d) are true *except* ________.
a) V_d relates the total amount of chemical in the body to the concentration of chemical in the plasma.

b) V_d is the apparent space into which an amount of chemical is distributed in the body to result in a given plasma concentration.
c) A chemical that usually remains in the plasma has a low V_d.
d) V_d will be low or a chemical with high affinity for tissues.
e) V_d can be used to estimate the amount of chemical in the body if the plasma concentration is known.

Q.33. Chemical clearance ________.
a) is independent of V_d
b) is unaffected by kidney failure
c) is indirectly proportional to V_d
d) is performed by multiple organs
e) is not appreciable in the GI tract

Q.34. Which of the following is *not* an advantage of a physiologically based toxicokinetic model?
a) Complex dosing regimens are easily accommodated.
b) The time course of distribution of chemicals to any organ is obtainable.
c) The effects of changing physiologic parameters on tissue concentrations can be estimated.
d) The rate constants are obtained from gathered data.
e) The same model can predict toxicokinetics of chemicals across species.

Q.35. Which of the following will not help to increase the flux of a xenobiotic across a biological membrane?
a) Decreased size
b) Decreased oil:water partition coefficient
c) Increased concentration gradient
d) Increased surface area
e) Decreased membrane thickness

Q.36. Which of the following statements is correct regarding diffusion-limited compartments?
a) Xenobiotic transport across the cell membrane is limited by the rate at which blood arrives at the tissue.
b) Diffusion-limited compartments are also referred to as flow-limited compartments.
c) Increased membrane thickness can cause diffusion-limited xenobiotic uptake.
d) Equilibrium between the extracellular and intracellular space is maintained by rapid exchange between the two compartments.
e) Diffusion of gases across the alveolar septa of a healthy lung is diffusion-limited.

Answers (Exercise 3)

1. b	5. b	9. e	13. d	17. c	21. d	25. a	29. d	33. d
2. c	6. d	10. d	14. b	18. a	22. c	26. b	30. c	34. d
3. c	7. d	11. b	15. a	19. b	23. b	27. a	31. a	35. b
4. e	8. a	12. d	16. d	20. d	24. d	28. c	32. d	36. c

Exercise 4

Q.1. The hepatic clearance of a drug with a high hepatic extraction ratio is largely dependent on ________.
a) drug protein binding
b) hepatic blood flow
c) drug-metabolizing enzyme activity
d) intestinal blood flow

Q.2. All of the following are of saturation kinetics with increasing dose *except* ________.
a) clearance must decrease
b) half-life can increase or decrease

 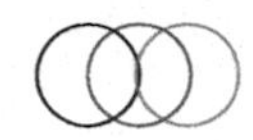

c) volume of distribution will decrease if there is saturation of serum protein binding
d) volume of distribution will decrease if there is saturation of tissue binding

Q.3. All of the following are of non-linear kinetics *except* ________.
a) ratio of metabolites will remain constant with change in dose
b) clearance will change with change in dose
c) AUC will not be dose proportional
d) decline of xenobiotic is non exponential

Q.4. All of the following are of first-order kinetics *except* ________.
a) steady-state concentration is proportional to rate of intake
b) rate of intake will not change time to steady state
c) half-life is inversely proportional to clearance
d) a change in half-life will not change time to steady-state

Q.5. All of the following are of first-order kinetics *except* ________.
a) The elimination rate constant increases with dose.
b) A semi-logarithmic plot of plasma concentration versus time yields a single straight line.
c) The concentration of xenobiotic in plasma decreases by a constant fraction per unit time.
d) The volume of distribution in independent of the dose.

Q.6. After ________, 93.8% of a dose of drug is eliminated.
a) 3 half-lives
b) 4 half-lives
c) 5 half-lives
d) 6 half-lives

Q.7. All of the following are components of the central compartment *except* ________.
a) liver
b) lungs
c) bone
d) kidney

Q.8. Which of the following has the largest value of distribution?
a) Chloroquine
b) Ethyl alcohol
c) Albumin
d) Ethylene glycol

Q.9. The common units used to express total clearance of a toxicant are ________.
a) mg/mL
b) mg/min
c) mL/min
d) mg/min/mL

Q.10. In first-order kinetics ________.
a) a constant amount of toxicant is removed over unit time
b) AUC is not proportional to dose
c) half-life changes with increasing dose
d) clearance, volume of distribution, and half-life do not change with dose

Q.11. Regarding the two-compartment model of classic toxicokinetics, which of the following is correct?
a) There is rapid equilibration of chemical between central and peripheral compartments.
b) The logarithm of plasma concentration versus time data yields a linear relationship.
c) There is more than one dispositional phase.
d) It is assumed that the concentration of a chemical is the same throughout the body.
e) It is ineffective in determining effective doses in toxicity studies.

Q.12. When calculating the fraction of a dose remaining in the body over time, which of the following factors need not be taken into consideration?
a) Half-life
b) Initial concentration
c) Time
d) Present concentration
e) Elimination rate constant

Q.13. All of the following statements regarding apparent volume of distribution (V_d) are true *except* ________.
a) V_d relates the total amount of chemical in the body to the concentration of chemical in the plasma.
b) V_d is the apparent space into which an amount of chemical is distributed in the body to result in a given plasma concentration.
c) A chemical that usually remains in the plasma has a low V_d.
d) V_d will be low for a chemical with high affinity for tissues.
e) V_d can be used to estimate the amount of chemical in the body if the plasma concentration is known.

Q.14. Chemical clearance ________.
a) is independent of V_d
b) is unaffected by kidney failure
c) is indirectly proportional to V_d
d) is performed by multiple organs.
e) is not appreciable in the GI tract.

Q.15. A chemical with which of the following half-lives (T1/2) will remain in the body for the longest period of time when given equal dosage of each?
a) T1/2 = 30 min
b) T1/2 = 1 day
c) T1/2 = 7 h
d) T1/2 = 120 s
e) T1/2 = 1 month

Q.16. With respect to first-order elimination, which of the following statements is *false*?
a) The rate of elimination is directly proportional to the amount of the chemical in the body.
b) A semi-logarithmic plot of plasma concentration versus time shows a linear relationship.
c) Half-life (T1/2) differs depending on the dose.
d) Clearance is dosage-independent.
e) The plasma concentration and tissue concentration decrease similarly with respect to the elimination rate constant.

Q.17. The toxicity of a chemical is dependent on the amount of chemical reaching the systemic circulation. Which of the following does *not* greatly influence systemic availability?
a) Absorption after oral dosing
b) Intestinal motility
c) Hepatic first-pass effect
d) Intestinal first-pass effect
e) Incorporation into micelles

Q.18. Which of the following is *not* an advantage of a physiologically based toxicokinetic model?
a) Complex dosing regimens are easily accommodated.
b) The time course of distribution of chemicals to any organ is obtainable.
c) The effects of changing physiologic parameters on tissue concentrations can be estimated.
d) The rate constants are obtained from gathered data.
e) The same model can predict toxicokinetics of chemicals across species.

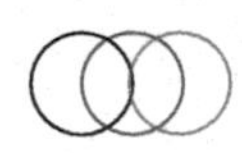

Q.19. Which of the following will *not* help to increase the flux of a xenobiotic across a biological membrane?
  a) Decreased size
  b) Decreased oil:water partition coefficient
  c) Increased concentration gradient
  d) Increased surface area
  e) Decreased membrane thickness

Q.20. Which of the following statements is true regarding diffusion-limited compartments?
  a) Xenobiotic transport across the cell membrane is limited by the rate at which blood arrives at the tissue.
  b) Diffusion-limited compartments are also referred to as flow-limited compartments.
  c) Increased membrane thickness can cause diffusion-limited xenobiotic uptake.
  d) Equilibrium between the extracellular and intracellular space is maintained by rapid exchange between the two compartments.
  e) Diffusion of gases across the alveolar septa of a healthy lung is diffusion-limited.

Q.21. If a chemical that is directly cytotoxic is detoxified by metabolism via the microsomal enzyme system and the activity of this system correlates with basal metabolic rate, which would you expect to be the species most sensitive to the chemical? All the species are given an equivalent dose on a weight basis.
  a) man
  b) mouse
  c) rat
  d) dog
  e) rabbit

Q.22. Metabolism of a foreign chemical will lead to ________.
  a) accumulation of the chemical in the tissues
  b) increased excretion in urine
  c) decreased toxicity
  d) altered chemical structure
  e) increased toxicity

Q.23. Cytochrome P450 is an enzyme that ________.
  a) is found in lysosomes
  b) is responsible for the conjugation of drugs
  c) is a central part of the drug-metabolizing system
  d) is one of the enzymes in the mitochondrial electron transport chain
  e) (c) and (d) are correct

Q.24. Phase 2 metabolism usually involves ________.
  a) microsomal enzymes
  b) decreasing the polarity of a chemical
  c) increasing the toxicity of compounds
  d) the addition of an endogenous moiety
  e) hydrolysis

Q.25. Glutathione is which of the following?
  a) A protein
  b) A tripeptide
  c) An enzyme involved in detoxication
  d) A substance found in the kidneys
  e) A vitamin

Q.26. Cytochrome P450 mainly catalyses the Phase 1 metabolism of chemicals
  a) if the statement is true
  b) if the statement is false

Q.27. The microsomal enzyme system is responsible for the metabolism of foreign compounds. Which of the following are essential aspects of this system?

1) Magnesium ions
2) The addition of two electrons
3) Molecular oxygen
4) The substrate is bound to an iron atom in the active site

Select the correct statement

a) 1, 2, and 3
b) 1 and 3
c) 2 and 4
d) only 4
e) all four

Answers (Exercise 4)

1. b	4. d	7. c	10. d	13. d	16. c	19. b	22. d	25. b
2. c	5. a	8. a	11. c	14. d	17. b	20. c	23. c	26. a
3. a	6. b	9. c	12. a	15. e	18. d	21. a	24. d	27. e

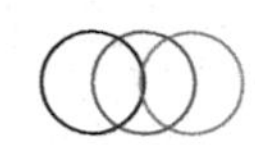

2.2 TRUE OR FALSE STATEMENTS

(Write T for True and F for False.)

Exercise 5

Q.1. A water-soluble drug will pass across muscle membranes faster than across brain membranes (assume permeability-rate limitations).
Q.2. A neutral, lipophilic drug is likely to be absorbed faster in the intestines than in the stomach. Remember that stomach and intestine differ in their properties.
Q.3. Lipophilic drugs are generally taken up fast by highly perfused organs.
Q.4. Ionized and lipophilic drugs are most likely to cross most membrane barriers.
Q.5. Drugs with a high tissue binding always have a large volume of distribution.
Q.6. Compared to skin, liver would have a higher rate of uptake of perfusion-limited lipophilic drugs due to its higher blood-flow rate.
Q.7. Distribution to a specific tissue for permeability-limited hydrophilic drugs depends on how much and how quickly the blood gets to the specific tissue.
Q.8. Perfusion limited distribution is a type of drug distribution into tissue that occurs when the drug is able to cross membranes easily.
Q.9. Assume two drugs (identical molecular weight, same dose given): one neutral drug (Drug A) and one acidic drug (pka = 7.4, Drug B). Drug A and the unionized form of drug B have the same partition coefficient. The fraction unbound in plasma and tissue is 0.5 for both drugs. Drug B will enter tissues somewhat slower than Drug A.
Q.10. A weak acid, whose unionized form shows a high partition coefficient, is likely to cross most membrane barriers.
Q.11. A volume of distribution of 41 L for a lipophilic drug suggests that the drug will not bind to tissue and plasma proteins.
Q.12. Transporters pumping the drug into the tissues are more active in babies.
Q.13. Compared to skin, liver would have a higher rate of uptake for small lipophilic drugs due to its higher blood-flow rate.
Q.14. Free drug concentrations are always the same in plasma and tissues, when the distribution occurs instantaneously.
Q.15. Enzyme induction affects the hepatic clearance of a low and high extraction drugs.
Q.16. Enzyme induction affects the oral bioavailability of high extraction drugs.
Q.17. A fast absorption might allow less-frequent dosing.
Q.18. A slower absorption might be advantageous for a drug with a narrow therapeutic window.
Q.19. Concentrations in plasma are of relevance for drug therapy as they generally correlate well with concentrations observed at the effect (target) site.
Q.20. When heparin is added to blood and the blood is centrifuged, the resulting supernatant is called serum.

Answers (Exercise 5)

1. T	2. T	3. T	4. F	5. F	6. T	7. F	8. T	9. T	10. T
11. F	12. T	13. T	14. T	15. F	16. T	17. F	18. T	19. T	20. F

Exercise 6

Assume a drug is substrate of a specific transport protein. Which of the following statements are true or false?

Q.1. Transporters do not use energy.
Q.2. Transporters only eliminate drugs from the body.

Q.3. Transporters are only present in liver and kidney.

Q.4. Transporters are saturable.

Q.5. Transporters work often in conjunction with enzymes.

Assume passive diffusion as the driving force for distribution. Which of the following statements are true or false?

Q.6. The Tl/2 of a drug eliminated through a zero-order process is a drug-specific constant.

Q.7. A lipophilic drug of low molecular weight cannot have a volume of distribution 1 that is smaller than V_T.

Q.8. The fraction of the drug being eliminated per hour is increasing in a first-order process.

Q.9. Two drugs that have similar elimination half-lives will have similar volumes of distributions.

Q.10. The same dose of a drug is given orally either as a solution or in form of as low dissolving crystal suspension. The solution will show higher maximum concentrations in plasma.

Answers (Exercise 6)

1. F 2. F 3. F 4. T 5. T 6. F 7. F 8. F 9. F 10. T

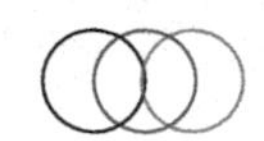

2.3 FILL IN THE BLANKS

Exercise 7

Q.1. The most common process of absorption of xenobiotics across the cell membrane is ________.
Q.2. The important route of excretion for xenobiotics is ________.
Q.3. The process of chemical transformation (conversion from one form to another) occurring in the body is known as ________.
Q.4. The major site for biotransformation of xenobiotics in body is ________.
Q.5. In a hepatocyte, metabolism of xenobiotics takes place in ________.
Q.6. The most important among microsomal enzymes is ________.
Q.7. The major biotransformation reaction occurring in Phase I is ________ and in Phase II is ________.
Q.8. Phase I oxidation reactions are mainly catalyzed by ________.
Q.9. All Phase II conjugation reactions are catalyzed by non-microsomal enzymes except for ________ which is catalyzed by microsomal enzymes.
Q.10. The ability of certain substances to increase the activity or synthesis of microsomal enzymes is known as ________.
Q.11. The metabolic reaction, which is deficient in dogs is ________; cats are deficient in ________ and pigs are deficient in ________ reactions of biotransformation.
Q.12. The process of conversion of non-toxic substance into a toxic metabolite due to biotransformation is known as ________.

Answers (Exercise 7)

1. Passive diffusion
2. Renal excretion
3. Biotransformation
4. Liver
5. Endoplasmic reticulum (ER) or microsomes
6. Mixed function oxidases (MFO)
7. Oxidation, conjugation
8. Microsomal enzymes (MFO)
9. Glucuronide conjugation
10. Induction
11. Acetylation; glucuronide conjugation and sulfation
12. Lethal synthesis

Exercise 8

Q.1. Drug or toxicant administration by any other route than the intravenous (IV) route is called ________.
Q.2. MEC is the ________ of drug in plasma required to produce the desirable pharmacological/therapeutic response.
Q.3. ________ is the maximal or peak pharmacological or toxic effect produced by the drug.
Q.4. First-order process is defined as a toxicokinetic process whose rate is ________ proportional to the concentration of the xenobiotic/chemical.
Q.5. Classic models ________ require information on tissue physiology or anatomic structure.
Q.6. The one-compartment open model is the simplest model that considers the ________ as a single.
Q.7. The central compartment consists of ________ and highly perfused organs.

Q.8. ________ may be defined as the time taken for the concentration of a compound/toxicant in plasma to decline by 1/2 or 50% of its initial value.

Q.9. When the rate of absorption of a compound is significantly slower than its rate of elimination from the body, it is known as ________.

Q.10. Dosing interval is the ________ between doses.

Answers (Exercise 8)

1. EV administration
2. Minimum concentration
3. Peak effect
4. Directly
5. Do not
6. Whole body
7. Blood
8. Half-life (T1/2)
9. Flip-flop kinetics
10. Time interval

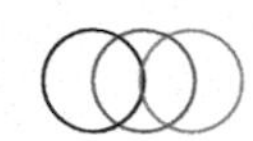

2.4 MATCH THE STATEMENTS

(Column A with Column B)

Exercise 9

Column A		Column B
Q.1.	metabolizing enzymes	a. p-aminobenzoic acid
Q.2.	dealkylation	b. oxidation
Q.3.	ethanol	c. permeability
Q.4.	procain	d. conjugation
Q.5.	Phase II reaction	e. liver
Q.6.	aspirin	h. acetaldehyde
Q.7.	transport	i. gastrointestinal
Q.8.	facilitated	j. salicylic acid
Q.9.	membrane	k. movement
Q.10.	oral	l. diffusion

Answers (Exercise 9)

1. e 2. b 3. h 4. a 5. d 6. j 7. k 8. l 9. c 10. i

Exercise 10

Column A		Column B
Q.1.	AUC	a. independent
Q.2.	onset of time	b. disposition
Q.3.	zero order	c. dermal
Q.4.	metabolism	d. total integrated area
Q.5.	first order	e. simplest
Q.6.	one compartment	f. start producing
Q.7.	flip-flop kinetics	g. directly proportional
Q.8.	PBTK	h. intramuscular
Q.9.	extra vascular	i. MSC
Q.10.	safe concentration	j. physiological

Answers (Exercise 10)

1. d 2. f 3. a 4. b 5. g 6. e 7. c 8. j 9. h 10. i

CHAPTER 3

MECHANISTIC TOXICOLOGY

3.1 MULTIPLE-CHOICE QUESTIONS

(Choose the most appropriate response.)

Exercise 1

Q.1. A possible reason for the selective embryo-fetal toxicity of DES is ________.
- a) higher concentration of free DES in embryo/fetal compared to adults
- b) binding to retinoic acid receptors
- c) lack of placental toxicant metabolism
- d) all of the above

Q.2. The liver and kidney are major target organs of toxicity because ________.
- a) they both receive a high percentage of cardiac output
- b) they both have substantial xenobiotic metabolizing capacity
- c) they both have transport systems that can concentrate xenobiotics
- d) all of the above

Q.3. Acyl glucuronides are particularly toxic to the liver because ________.
- a) they selectively interact with macrophages releasing active oxygen
- b) active transport systems in the hepatocyte and bile duct system can greatly up concentrate them
- c) they are resistant to glucuronidase
- d) they are suitable inhibitors of UGT2B7

Q.4. The selective renal toxicity of cephaloridine over cephalothin is due to ________.
- a) selective uptake by the organic cation transporter
- b) selective inhibition of P-glycoprotein
- c) selective uptake by the organic anion transporter
- d) significantly less plasma protein binding of cephaloridine

Q.5. All of the following are true of alpha-amanitin *except* ________.
- a) It is less orally available than phalloidin.
- b) It inhibits RNA polymerase II.
- c) It is transported into the hepatocyte by a bile-acid transporter.
- d) It is a mushroom toxin.

Q.6. All of the following are true of the toxic mechanism of paraquat *except* ________.
- a) Lungs accumulate paraquat in an energy-dependent manner.
- b) Its energy into the lungs is assumed to be via the polyamine transport system.
- c) Similar molecules with smaller distances between nitrogen atoms do not enter lungs as readily.
- d) Cytotoxicity to alveolar cells is caused by interference with calcium channels.

Q.7. Enzyme induction of phenobarbital is mediated through ________.
- a) aryl hydrocarbon receptor
- b) PPAR-alpha receptor
- c) constitutively active receptor (CAR)
- d) estrogen receptor

Q.8. CAR is down regulated by ________.
  a) hypericum extracts
  b) acetaminophen
  c) aspirin
  d) proinflammatory cytokines

Q.9. The pregnane X receptor ________.
  a) is a cytosolic receptor
  b) is involved in induction of CYP3A4
  c) is primarily expressed in skin
  d) all of the above

Q.10. Downregulation of receptors is due to continuous exposure of ________.
  a) antagonist
  b) agonist
  c) inverse agonist
  d) all the above

Q.11. Amphipathic xenobiotics that can become trapped in lysosomes and cause phospholipidosis include all of the following *except* ________.
  a) ethylene glycol
  b) amiodarone
  c) amitriptyline
  d) fluoxetine

Q.12. Which of the following parent toxicant-electrophilic metabolite pairs is incorrect?
  a) halothane–phosgene
  b) bromobenzene–bromobenzene 3, 4 oxide
  c) benzene–muconic aldehyde
  d) allyl alcohol–acrolein

Q.13. All of the following are capable of accepting the electrons from reductases and forming radicals *except* ________.
  a) paraquat
  b) doxorubicin
  c) *n*-hexane
  d) nitrofurantoin

Q.14. An example of the formation of an electrophilic toxicant from an inorganic chemical is ________.
  a) CO to CO_2
  b) AsO_4
  c) NO to NO_2
  d) hydroxide ion to water

Q.15. The general mechanism for detoxification of electrophiles is ________.
  a) conjugation with glucuronic acid
  b) conjugation with acetyl CoA
  c) conjugation with glutathione
  d) conjugation with sulfate

Q.16. The most common nucleophilic detoxification reaction that amines undergo is ________.
  a) acetylation
  b) sulfation
  c) methylation
  d) amino-acid conjugation

Q.17. Detoxification mechanisms fail because ________.
  a) toxicants may overwhelm the detoxification process
  b) a reactive toxicant may inactivate a detoxicating enzyme.

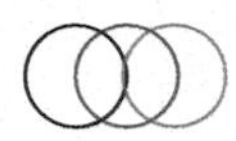

c) detoxication may produce a toxic by-product
d) all of the above

Q.18. To which receptors can agonist bind but are unable to elicit chemical response?
a) Spare receptors
b) Orphan receptors
c) Silent receptors
d) None

Q.19. The hydroxyl radical can be produced by all of the following *except* ________.
a) the action of nitric oxide synthetase on water
b) interaction of ionizing radiation and water
c) reductive homolytic fission of hydrogen peroxide
d) interaction of silica with surface iron ions in lung tissue

Q.20. If an electrophile is covalently bound to a protein that does not play a critical function, the result is considered a ________.
a) toxication reaction
b) detoxication reaction
c) MNA adduct formation
d) Fenton reaction

Q.21. Which of the following receptor–exogenous ligand pairs is incorrect?
a) Estrogen receptor–zearalenone
b) Glucocorticoid receptor–dexamethsasone
c) Aryl hydrocarbon receptor–rifampicin
d) Peroxisome proliferator-activated receptor–clofibrate

Q.22. Which of the following receptor–agonist pairs incorrect?
a) Glutamate receptor–kainate
b) Glycine receptor–strychnine
c) GABA (A) receptor–muscimol
d) Opioid receptor–meperidine

Q.23. Which of the following receptor–antagonist pairs is incorrect?
a) Adrenergic beta I receptor–metoprolol
b) Serotonin (2) receptor–ketanserin
c) Glutamate receptor–ketamine
d) GABA (A) receptor–ivermectin

Q.24. Clonidine overdose mimics poisoning with ________.
a) morphine
b) cocaine
c) phencyclidine
d) amphetamine

Q.25. All of the following act as inhibitors of the citric acid cycle *except* ________.
a) 4-pentenoic acid
b) fluoroacetate
c) DCVC (S-(1,2-dichlorovinyl)-L-cysteine)
d) malonate

Q.26. All of the following are inhibitors of ADP phosphorylation *except* ________.
a) oligomycin
b) DDT
c) ethanol
d) *N*-ethylmaleimide

Q.27. All of the following cause calcium influx into the cytoplasm *except* ________.
a) capsaicin
b) formate

c) domoate
d) amphotericin B

Q.28. All of the following inhibit calcium export from the cytoplasm *except* ________.
a) vanadate
b) methylmercury
c) bromobenzene
d) carbon tetrachloride

Q.29. The hydroxyl radical is enzymatically detoxified by ________.
a) catalase
b) glutathione peroxide
c) glutathione reductase
d) none of the above

Q.30. Which of the following is regarding cell death?
a) Necrosis requires ATP
b) Release of cytochrome c usually triggers necrosis
c) Toxicants at low doses usually cause apoptosis and necrosis at higher doses
d) Apoptosis is never a desirable effect

Q.31. The severity of a toxin depends, in large part, on the concentration of the toxin at its site of action. Which of the of following will decrease the amount of toxin reaching its site of action?
a) Absorption across the skin
b) Excretion via the kidneys
c) Toxication
d) Reabsorption across the intestinal mucosa
e) Discontinuous endothelial cells of hepatic sinusoids

Q.32. Toxication (or metabolic activation) is the biotransformation of a toxin to a more toxic and reactive species. Which of the following is not a reactive chemical species commonly formed by toxication?
a) Electrophiles
b) Nucleophiles
c) Superoxide anions
d) Hydroxy radicals
e) Hydrophilic organic acids

Q.33. Which of the following is not an important step in the detoxication of chemicals?
a) Formation of redox active reactants
b) Reduction of hydrogen peroxide by glutathione peroxidase
c) Formation of hydrogen peroxide by superoxide dismutase
d) Reduction of glutathione disulfide (GSSG) by glutathione reductase (GR)
e) Conversion of hydrogen peroxide to water and molecular oxygen by catalase

Q.34. Regarding the interaction of the ultimate toxicant with its target molecule, which of the following is false?
a) Toxins often oxidize or reduce their target molecules, resulting in the formation of a harmful by-product.
b) The covalent binding of a toxin with its target molecule permanently alters the target's function.
c) The non-covalent binding of a toxin to an ion channel irreversibly inhibits ion flux through the channel.
d) Abstraction of hydrogen atoms from endogenous compounds by free radicals can result in the formation of DNA adducts.
e) Several toxins can act enzymatically on their specific target proteins.

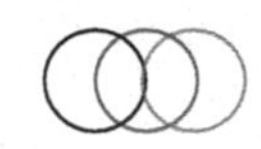

Q.35. All of the following are common effects of toxicants on target molecules *except* ________.
   a) blockage of neurotransmitter receptors
   b) interference with DNA replication due to adduct formation
   c) cross-linking of endogenous molecules
   d) opening of ion channels
   e) mounting of an immune response

Q.36. Which of the following proteins functions to prevent the progression of the cell cycle?
   a) NF-κB
   b) MAPK
   c) CREB
   d) c-Myc
   e) IκB

Q.37. Which of the following would have the largest negative impact on intracellular ATP levels?
   a) Moderately decreased caloric intake
   b) Interference with electron delivery to the electron transport chain
   c) Inability to harvest ATP from glycolysis
   d) Increased synthesis of biomolecules
   e) Active cell division

Q.38. What happens when a toxin induces elevation of cytoplasmic calcium levels?
   a) Mitochondria uptake of calcium dissipates the electrochemical gradient needed to synthesize ATP.
   b) Formation of actin filaments increases the strength and integrity of the cytoskeleton.
   c) It decreases the activity of intracellular proteases, nucleases, and phospholipases.
   d) The cell becomes dormant until the calcium is actively pumped from the cell.
   e) The generation of reactive oxygen species slows because of the calcium-induced decrease in activity of the TCA cycle.

Q.39. Cytochrome c is an important molecule in initiating apoptosis in cells. All of the following regarding cytochrome c are correct *except* ________.
   a) The release of cytochrome c into the cytoplasm is an important step in apoptosis initiation.
   b) The loss of cytochrome c from the electron transport chain blocks ATP synthesis by oxidative phosphorylation.
   c) Loss of cytochrome c from the inner mitochondria membrane results in increased formation of reactive oxygen species.
   d) Bax proteins mediate cytochrome c release.
   e) Caspases are proteases that increase cytoplasmic levels of cytochrome c

Q.40. All of the following regarding DNA repair are true *except* ________.
   a) In a lesion that does not cause a major distortion of the double helix, the incorrect base is cleaved and the correct base is inserted in its place.
   b) Base excision repair and nucleotide excision repair are both dependent on a DNA polymerase and a DNA ligase.
   c) In nucleotide excision repair, only the adduct is cleaved, and the gap is then filled by DNA polymerase.
   d) Pyrimidine dimers can be cleaved and repaired directly by DNA photolyase.
   e) Recombinational repair requires that a sister strand serve as a template to fill in missing nucleotides.

Q.41. Which kind of cells are the primary targets of the AIDS virus?
   a) Cytotoxic (killer) cell
   b) Helper T-cells (e.g., CD4)
   c) Memory cells
   d) Suppressor T-cells
   e) Delayed hypersensitivity T-cells

Q.42. The largest percent of antibodies belong to the ________ class.
a) IgG
b) IgE
c) IgM
d) IgA
e) IgD

Q.43. Which of the following hypersensitivity reactions is most often seen in transfusion reactions?
a) Cytotoxic
b) Cell-mediated
c) Immune complex
d) Anaphylactic
e) None of the above

Q.44. What kind of hypersensitivity is associated with asthma?
a) Cytotoxic
b) Cell-mediated
c) Immune complex
d) Anaphylactic
e) None of the above

Q.45. What kind of the following interactions is characteristic of that for caffeine and sleeping pills?
a) Additive
b) Synergistic
c) Antagonistic
d) None of the above
e) They don't interact with each other.

Q.46. Yu-Cheng disease in Taiwan is due to the toxic effect of ________.
a) lead
b) PCBs
c) dioxin
d) asbestos
e) mercury

Q.47. When were both federal regulatory and legislative efforts begun to reduce lead hazards, including the limitation of lead in paint and gasoline?
a) 1970s
b) 1980s
c) 1990s
d) 2000s
e) 2015

Q.48. Cytochrome P450 activity can be affected by ________.
a) foods
b) social habits
c) thyroid disease
d) all of the above

Q.49. An example of oxidative desulfuration is ________.
a) parathion to paraoxon
b) imipramine to desipramine
c) codeine to morphine
d) enalapril to enalaprilat

Q.50. CYP3A7 is present mostly in ________.
a) adult human liver
b) fetal human liver

 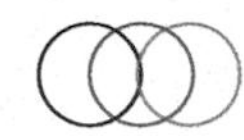

c) rodent liver
d) human lymphoma cells

Q.51. Paracetamol is an analgesic drug that may cause liver damage after overdoses. This is the result of ________.
a) depletion of body stores of sulfate
b) inhibition of cytochrome P450
c) production of a glutathione conjugate
d) metabolic activation by the microsomal enzymes
e) biliary excretion and metabolism by the gut bacteria

Q.52. Indicate which of the following chemicals is metabolized to oxalic acid.
a) Methanol
b) Fluoroacetic acid
c) Ethylene glycol
d) Naphthalene
e) Benzene

Q.53. Which of the following is the most important in determining the extent of toxicity of a chemical?
a) Chemical structure
b) Dose
c) Metabolism of the compound
d) Excretion of the compound
e) Metabolic detoxication of the compound

Q.54. If an animal is exposed to the solvent benzene, which of the following will be observed?
a) Benzene will be mainly excreted unchanged in the expired air.
b) Benzene will be metabolized mainly to phenol and excreted as water soluble conjugated metabolites.
c) Benzene will be entirely localized in body fat.
d) Benzene will be excreted into the bile.
e) None of the above.

Q.55. Ethanol is used as an antidote for the treatment of ethylene-glycol poisoning because it ________.
a) facilitates the excretion of ethylene glycol
b) blocks the metabolism of ethylene glycol
c) increases the detoxication of ethylene glycol
d) chelates ethylene glycol
e) none of the above

Q.56. In patients poisoned with aspirin, the symptoms include ________.
1) respiratory alkalosis
2) metabolic alkalosis
3) metabolic acidosis
4) respiratory acidosis

Select the correct statement
a) 1, 2, and 3
b) 1 and 3
c) 2 and 4
d) only 4
e) all four

Q.57. Isoniazid toxicity to the peripheral nerves ________.
1) is due to the parent drug
2) only occurs in rapid acetylators

3) is the result of depletion of pyridoxal
4) is possibly due to a reactive metabolite
Select the correct statement
a) 1, 2, and 3
b) 1 and 3
c) 2 and 4
d) only 4
e) all four

Q.58. The hepatic toxicity of bromobenzene:
1) is due to a glutathione conjugate
2) is due to inhibition of cytochrome P450
3) is decreased by phenobarbital pretreatment
4) is due to a reactive-epoxide intermediate
Select the correct statement
a) 1, 2, and 3
b) 1 and 3
c) 2 and 4
d) only 4
e) all four

Answers (Exercise 1)

1. a	6. d	11. a	16. a	21. c	26. c	31. b	36. e	41. b	46. b	51. d	56. b
2. d	7. c	12. a	17. d	22. b	27. b	32. e	37. b	42. a	47. a	52. c	57. b
3. b	8. d	13. c	18. c	23. d	28. b	33. a	38. a	43. a	48. d	53. b	58. d
4. c	9. b	14. b	19. a	24. a	29. d	34. c	39. e	44. d	49. a	54. a	
5. a	10. b	15. c	20. b	25. a	30. c	35. d	40. c	45. c	50. b	55. b	

Exercise 2

Q.1. Major target molecules for toxicants include all of the following *except* ________.
a) proteins
b) vitamins
c) DNA
d) lipids

Q.2. All of the following toxins act by enzymatic reaction *except* ________.
a) ricin
b) anthrax
c) tetrodotoxin
d) botulinum

Q.3. Apoptotic pathways can be initiated by ________.
a) DNA damage
b) mitochondrial insult
c) death–receptor stimulation
d) all of the above

Q.4. The enzyme that repairs oxidized protein thiols is called ________.
a) HMG-coenzyme A reductase
b) adenyl cyclase
c) phospholipase
d) none of the above

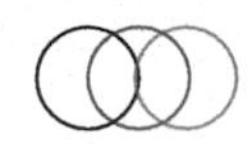

Q.5. The mechanism of action for a bleomycin-induced lung injury is presumed to include ________.
- a) DNA-adduct formation
- b) generation of reactive oxygen species
- c) inhibition of cytochrome oxidase
- d) none of the above

Q.6. All of the following are true of oxidative DNA damage *except* ________.
- a) Mitochondrial DNA is much more resistant to damage that nuclear DNA.
- b) B-hydroxy-deoxyguanosine in the urine is a marker.
- c) It can lead to base pair transversions.
- d) It can lead to a point mutation.

Q.7. An example of denatured protein is ________.
- a) golgi complex
- b) micronuclei
- c) heinz body
- d) histone

Q.8. An important feature of lipid peroxidation is ________.
- a) it cannot be blocked by antioxidants
- b) damage can be propagated in a chain reaction-like manner
- c) it never involves the Fenton reaction
- d) the end products are different from the end products of the reaction of lipids with ozone

Q.9. All of the following are true regarding mechanisms of immune-system toxicology *except* ________.
- a) TCDD-induced thymic atrophy may be mediated by the aryl hydrocarbon receptor
- b) The addition of a happen to a protein may cause a conformational change that displays previously hidden antigenic regions.
- c) Oral exposure of a xenobiotic is associated with a much greater chance of an immune reaction than by other routes.
- d) The 'danger hypothesis' refers to a break in immune tolerance to an antigen triggered by signals initiated by cellular or systemic stress.

Q.10. Which statement is regarding the PPAR alpha-receptor?
- a) Stimulation causes peroxisome proliferation in humans.
- b) They are present in adipose tissues.
- c) They are involved in fatty-acid beta oxidation.
- d) Thiazolidinediones act as ligands.

Q.11. Alterations in retinoic acid receptor (RAR) function have been associated with ________.
- a) cardiac and vascular toxicity
- b) CNS and peripheral nerve toxicity
- c) hepatic and renal toxicity
- d) embryo and testicular toxicity

Q.12. Proteins present in brown adipose tissue act mechanically like ________.
- a) pentachlorophenol
- b) rotenone
- c) cyanide
- d) doxorubicin

Q.13. Which of the following regarding glutathione is false?
- a) It can act non-enzymatically as a radical scavenger.
- b) It is a dipeptide.
- c) Its levels can be upregulated in response to a need.
- d) It is a substrate for glutathione peroxidase.

Q.14. All of the following are associated with necrosis *except* ________.
a) requirement of ATP
b) cell swelling
c) association with an inflammatory response
d) initiation by plasma membrane permeability changes

Q.15. Another name for apoptosis is ________.
a) passive cell death
b) accidental cell death
c) programmed cell death
d) immune cell death

Q.16. Apoptosis can serve as a tissue repair process in a number of cell types. In which of the following cell types would this be a plausible mechanism of tissue repair?
a) Female germ cells
b) Gastrointestinal epithelium
c) Neurons
d) Retina ganglion cells
e) Cardiac muscle cells

Q.17. A mitochondrial factor involved in the process of apoptosis is ________.
a) cytochrome a3
b) cytochrome b6
c) cytochrome c
d) cytochrome a1

Q.18. A class of toxicants that can induce apoptosis in certain cancer cells is ________.
a) barbiturates
b) digitalis glycosides
c) protein pump inhibitors
d) COX-2 inhibitors

Q.19. Which of the following signal-transduction pathway-effect pairs is correct?
a) ras/ERK–suppression of apoptosis
b) JNK–mediate apoptosis
c) p38–production of inflammatory cytokines
d) All of the above

Q.20. A toxicant that works by disrupting mitosis is ________.
a) paclitaxel
b) doxorubicin
c) methotrexate
d) cyclophosphamide

Q.21. Galactosamine causes ________.
a) proximal tubular damage
b) focal liver necrosis
c) peripheral neuropathy
d) all of the above

Q.22. Practolol was withdrawn from use because of ________.
a) unpredictable beta blockade
b) rebound hypertension
c) teratogenicity
d) oculomucocutaneous syndrome

Q.23. All of the following can inhibit the opening of the mitochondrial permeability transition pore (mPT) *except* ________.
a) cyclosporine A
b) hydrophobic bile ducts

 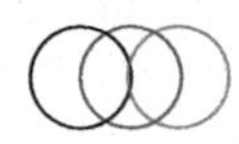

c) L-deprenyl
d) bongkrekic acid

Q.24. Microcystins target ________.
a) adenyl cyclase
b) phosphodiesterase
c) protein phosphatase
d) guanyl cyclase

Q.25. All of the following are true of cytokines *except* ________.
a) short half-life
b) act locally
c) produced in specialized organs
d) have complex interactions

Q.26. Molecular chaperones ________.
a) repair denatured proteins
b) repair damaged DNA
c) repair damaged transfer RNA
d) act as catalyst for new protein synthesis

Q.27. Lipid repair ________.
a) cannot occur
b) requires NADPH
c) may involve catalase
d) none of the above

Q.28. In peripheral neurons with damaged axons, repair ________.
a) can only occur until approximately age 3
b) follows the same mechanism as CNS repair
c) requires activated neutrophils and astrocytes
d) requires macrophages and Schwann cells

Q.29. Apoptosis is most advantageous in ________.
a) neoplastic prostate cells
b) female germ cells
c) cardiac myocytes
d) CNS neurons

Q.30. The principal factor leading to fibrogenesis is ________.
a) IL-8
b) INF alpha
c) TGF beta
d) IL-1

Q.31. Fibrosis is harmful because ________.
a) it may compress blood cells
b) it may contribute to tissue malnutrition
c) it may interfere with mechanical organ function.
d) all of the above

Q.32. All of the following are true regarding idiosyncratic toxicant reactions *except* ________.
a) They are rare.
b) They are predictable from the pharmacology of the drug.
c) The reaction can be dose dependent.
d) They involve genetic or acquired factors that increase susceptibility.

Q.33. All of the following would indicate initiation of a cellular stress response *except* ________.
a) Down regulation of DNA repair enzymes
b) Induction of apoptosis
c) Up regulation of antioxidant mechanisms
d) Stimulation of an immune response

Q.34. All of the following contribute to organ-selective toxicity *except* ________.
a) organ-selective uptake
b) tissue-specific expression of transcription factors
c) number of chromosomes in nucleus
d) tissue-specific receptors

Q.35. The nephrotoxic effect of mercury on the kidney is thought to be mediated by ________.
a) blocking the effect of ADH on the collecting duct
b) interfering with ionic charges on the glomerulus
c) a dicysteinyl-mercury complex mimicking endogenous cystine
d) interfering with chloride transport in the loop of Henle

Q.36. An example of a soft neutrophile is ________.
a) sulfur in glutathione
b) phosphate oxygen in nucleic acids
c) mercuric ion
d) carboxylate anion

Q.37. The Fenton reaction produces ________.
a) phosgene from chloroform
b) formic acid from formaldehyde
c) nitrogen dioxide from ozone and nitrogen
d) hydroxyl radical and hydroxyl ions from hydrogen peroxide

Q.38. Which is the most likely process of absorption for amino acids?
a) Diffusion
b) Facilitated diffusion
c) Active transport
d) Endocytosis
e) None of the above

Q.39. Which of the following sites in the respiratory system is the most likely place for the carbon dioxide and oxygen to exchange in the blood?
a) Nose
b) Pharynx
c) Larynx
d) Trachea
e) Alveoli

Q.40. Which of the following processes, when prolonged and severe, can be life threatening, such as in asthmatic attacks?
a) Mucociliary streaming
b) Coughing
c) Sneezing
d) Bronchoconstriction
e) None of the above

Q.41. What is the best estimate for the area of skin coverage in the average adult human?
a) 2500 in^2
b) 3000 in^2
c) 3500 in^2
d) 4000 in^2
e) 4500 in^2

Q.42. By what absorptive process does hexane pass through the skin?
a) Passive diffusion
b) Facilitated diffusion
c) Active transport
d) Endocytosis
e) None of the above

Q.43. How long is the average adult human gastrointestinal tract?
a) 20 feet
b) 22 feet
c) 30 feet
d) 41 feet
e) 52 feet

Q.44. Where is the most likely site for the absorption of toxic agents in the gastrointestinal tract?
a) Between the stomach and the upper portion of the intestine
b) Stomach
c) Small intestine
d) Large intestine
e) The lower portion of large intestine

Q.45. What is the mechanism for the harmful effects of CO (carbon monoxide)?
a) Interfere with or block the active sites of some important enzymes
b) Direct chemical combination with a cell constituent
c) Secondary action as a result of its presence in the system
d) Compete with the cofactors for a site on an important enzyme
e) None of the above

Q.46. What is the major organ responsible for detoxification in the body?
a) Lung
b) Intestines
c) Kidney
d) Liver
e) Skin

Q.47. What is the major toxic mechanism for hydrogen cyanide?
a) Interfere with or block the active sites of the enzyme
b) Inactivate or remove the cofactor
c) Compete with the cofactor for a site on the enzyme
d) Altering enzyme structure directly thereby changing the specific three-dimensional nature of the active site
e) None of the above

Q.48. What is the major mechanism for toxicity of dithiocarbamate during alcohol consumption?
a) Interfere with or block the active sites of the enzyme
b) Inactivate or remove the cofactor
c) Compete with the cofactor for a site on the enzyme
d) Altering enzyme structure directly thereby changing the specific three-dimensional nature of the active site
e) None of the above

Q.49. Which of the following substances can cause a syndrome in infants referred to as 'blue baby'?
a) Carbon monoxide
b) Chlorine gas
c) Ozone
d) Sulfuric oxides
e) Nitrogen compounds

Q.50. Which of the following refers to a substance that is attached to an antigen and promotes an antigenic response?
a) Light chain
b) Heavy chain
c) Leukocyte
d) Helper cell
e) Hapten

Q.51. What form of ionizing radiation has the shortest range (i.e., travels the shortest distance in tissue) for the same initial energy?
a) Alpha particle
b) Beta particle
c) Gamma ray
d) X-ray

Q.52. What property of beryllium-containing compounds is the reason they are not readily absorbed from the gastrointestinal tract?
a) They form insoluble phosphate precipitates at the pH of the intestinal tract.
b) They are chelated by bile salts in the small intestine.
c) Their size prevents passage through cell membranes.
d) They are converted to an oxide form.

Q.53. What is the mechanistic basis for cyclosporin-A-induced cholestasis?
a) Interaction with the cytoplasmic receptor, cyclophilin
b) Inhibition of transcription of message (mRNA) for critical cytokines
c) Induction of programmed cell death in centrilobular hepatocytes
d) Inhibition of ATP-dependent transporter(s) in the bile canalicular membrane

Q.54. What is the mechanism of neurotoxicity for strychnine?
a) Glycine receptor antagonist
b) GABA uptake inhibitor
c) Glutamate receptor agonist
d) Blockade of muscarinic cholinoceptors

Q.55. What condition will increase the transport of a substance across the blood–brain barrier?
a) High ionization
b) Strong binding to lipoproteins
c) Strong binding to plasma proteins
d) Cysteine binding

Answers (Exercise 2)

1. b	7. c	13. b	19. d	25. c	31. d	37. d	43. c	49. e	54. a
2. c	8. b	14. a	20. a	26. a	32. b	38. b	44. a	50. e	55. d
3. d	9. c	15. c	21. b	27. b	33. a	39. e	45. b	51. a	
4. d	10. c	16. b	22. d	28. d	34. c	40. d	46. d	52. a	
5. b	11. d	17. c	23. b	29. a	35. c	41. b	47. b	53. d	
6. a	12. a	18. d	24. c	30. c	36. a	42. a	48. b		

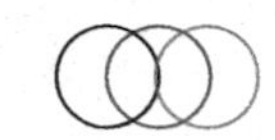

3.2 TRUE OR FALSE STATEMENTS

(Write T for True or F for False)

Exercise 3

Q.1. When lead covalently bonds to an enzyme, its inhibition of enzymes is considered to be irreversible.
Q.2. The exposure of allergens can trigger a diminished immune response in some people.
Q.3. Chemical pollutants such as ozone can depress the immune response by inactivating alveolar macrophages.
Q.4. B cells are the principle agents in cell-mediated immunity.
Q.5. Humoral immune responses are characterized by subcutaneous bleeding.
Q.6. The environmental pollutants, such as ozone and fine particulates, contribute to the significant rise in the numbers and severity of asthma cases.
Q.7. If absorbed, lead tends to be stored mostly in fatty tissue.
Q.8. Dioxin is considered to be one of the most toxic natural chemicals.
Q.9. The EPA has listed 20 μg/dl as the maximum acceptable blood lead level for fetuses and young children.
Q.10. Lead may impair fertility in both men and women when blood lead levels approach 50 μg/dl.
Q.11. Lead has been recognized as a hazard since early civilization when it was used to store wine, to pipe water, and even as vessels in which to cook food.
Q.12. Once a potential toxic substance goes into our society, it automatically produces an adverse effect.
Q.13. External respiration refers to the exchange of gases between blood and individual cells.
Q.14. Sulfur oxides tend to reach deep into lung tissue while nitrogen dioxides tend to act in the upper moist airways of the respiratory tree.
Q.15. The skin is the body's largest organ and consists of many interconnected tissues.
Q.16. Epidermis is the outer, thinner layer of the skin, and dermis is the inner and much thicker layer of the skin.
Q.17. The gastrointestinal tract is a major route of absorption for many toxic agents including mercury, lead, and cadmium.
Q.18. A toxin can produce a harmful effect upon an organ only by stimulating the normal metabolic actions of that particular organ.
Q.19. Many enzymes require a non-protein component called an apoenzyme and a protein component called a cofactor to become active.
Q.20. Cadmium and beryllium are believed to inactivate enzymes by blocking the sites on the enzyme where such cofactors as iron normally attach.

Answers (Exercise 3)

1. T	3. T	5. T	7. F	9. F	11. T	13 F	15. T	17. T	19. F
2. F	4. F	6. T	8. F	10. T	12. F	14. F	16. T	18. F	20. F

3.3 FILL IN THE BLANKS

Exercise 4

Q.1. Non-specific acting toxicants are those that produce ________.

Q.2. Uncouplers of oxidative phosphorylation uncouple the two processes that occur in ________, i.e., electron transfer and adenosine triphosphate (ATP) production.

Q.3. Specific acting toxicants are those that are non-narcotic and that produce a specific action at a specific ________.

Q.4. Irritants are chemicals that cause ________ on living tissue by chemical action at the site of contact.

Q.5. Chemicals that cause DNA adducts can lead to ________, which can activate cell-death pathways.

Q.6. Chemicals that cause protein adducts can lead to ________, which can activate cell-death pathways.

Q.7. Necrosis can lead to unprogrammed ________.

Q.8. Necrosis is caused by factors external to the cell or tissue, such as ________, ________.

Q.9. Apoptosis is the term used to describe generally the normal death of the cell in ________.

Q.10. Toxic chemicals can produce ________ or ________ bodily injury.

Answers (Exercise 4)

1. Narcosis
2. Oxidative phosphorylation
3. Specific target site
4. Inflammation
5. DNA mutations
6. Protein dysfunction
7. Cell death
8. Infection, toxins, or trauma
9. Living organisms
10. Reversible, irreversible

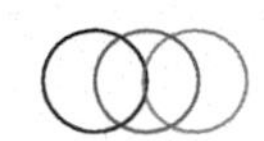

3.4 MATCH THE STATEMENTS

(Column A with Column B)

Exercise 5

Column A		Column B	
Q.1.	muscimol	a.	serotonin agonist
Q.2.	benzodiazepines	b.	prevents vesicle dopamine uptake
Q.3.	clonidine	c.	inhibits norepinephrine uptake
Q.4.	baclofen	d.	direct nicotine antagonist
Q.5.	bicuculline	e.	direct dopamine agonist
Q.6.	theophylline	f.	direct serotonin antagonist/glycine uptake inhibitor
Q.7.	nicotine	g.	direct GABA (A) agonist
Q.8.	clozapine	h.	indirect GABA (A) agonist
Q.9.	tecadenoson	i.	alpha-2 adrenoceptor agonist
Q.10.	yohimbine	j.	dopamine antagonist
Q.11.	cocaine	k.	direct GABA (B) agonist
Q.12.	alpha-bungarotoxin	l.	adenosine antagonist
Q.13.	botulinum toxin	m.	direct adenosine agonist
Q.14.	bromocriptine	n.	alpha-2 adrenoceptor antagonist
Q.15.	haloperidol	o.	direct GABA (A) antagonist
Q.16.	reserpine	p.	agonist at neuromuscular junction
Q.17.	ergonovine	q.	inhibits acetylcholine release

Answers (Exercise 5)

1. g 2. h 3. i 4. k 5. o 6. i 7. p 8. f 9. m 10. n
11. c 12. d 13. q 14. e 15. j 16. b 17. a

SECTION 2
ORGAN TOXICITY

Chapter 4

TARGET ORGAN TOXICITY

4.1 MULTIPLE-CHOICE QUESTIONS

(Choose the most appropriate response.)

Exercise 1

Q.1. The kidney is responsible for all of the following *except* ________.
  a) synthesis of renin
  b) acid–base balance
  c) reabsorption o electrolytes
  d) regulation of extracellular fluid
  e) release of angiotensin

Q.2. Which of the following does *not* contribute to filtrate formation in the nephron?
  a) Capillary hydrostatic pressure
  b) Positive charge of glomerular capillary wall
  c) Hydraulic permeability of glomerular capillary wall
  d) Colloid oncotic pressure
  e) Size of filtration slits

Q.3. Which of the following is *not* a characteristic of the loop of Henle?
  a) There is reabsorption of filtered Na^+ and K^+.
  b) Tubular fluid in the thin descending limb is iso-osmotic to the renal interstitium.
  c) Water is freely permeable in the thin ascending limb.
  d) Na^+ an Cl^- are reabsorbed in the thin ascending limb.
  e) The thick ascending limb is impermeable to water.

Q.4. Although the kidneys constitute 0.5% of total body mass, approximately how much of the resting cardiac output do they receive?
  a) 0.5%–1%
  b) 5%
  c) 10%
  d) 20%–25%
  e) 50%–60%

Q.5. Which of the following is most likely to occur after a toxic insult to the kidney?
  a) GFR will decrease in the unaffected kidney.
  b) Tight junction integrity will increase in the nephron.
  c) The unaffected cells will undergo atrophy and proliferation.
  d) Clinical tests will likely show normal renal function.
  e) Glomerulotubular balance is lost.

Q.6. Chronic renal failure does not typically result in ________.
  a) decrease in GFR of viable nephrons
  b) glomerulosclerosis
  c) tubular atrophy
  d) increase glomerular pressures
  e) altered capillary permeability

Q.7. All of the following statements regarding toxicity to the kidney are true *except* ________.
  a) Concentration of toxins in tubular fluid increase the likelihood that the toxin will diffuse into tubular cells.

b) Drugs in the systemic circulation are delivered to the kidneys at relatively high amounts.
c) The distal convolute tubule is the most common site of toxicant-induced renal injury.
d) Immune complex deposition within the glomeruli can lead to glomerulonephritis.
e) Antibiotics and /or anti-fungal drugs affect the functioning of the nephron at multiple locations.

Q.8. Which of the following test results is *not* correctly paired with the underlying kidney problem?
a) Increase urine volume–effect in ADH synthesis
b) Glucosuria–effect in reabsorption in the proximal convolute tubule
c) Proteinuria–glomerular damage
d) Proteinuria–proximal tubular injury
e) Brush-border enzymuria–glomerulonephritis

Q.9. Renal cell injury is *not* commonly mediated by which of the following mechanisms?
a) Loss of membrane integrity
b) Impairment of mitochondrial function
c) Increase cytosolic Ca^{2+} concentration
d) Increase Na^+, K^+-ATPase activity
e) Caspase activation

Q.10. Which of the following statements is FALSE with respect to nephrotoxicants?
a) Mercury poisoning can lead to proximal tubular necrosis an acute renal failure.
b) Cisplatin may cause nephrotoxicity because of its ability to inhibit DNA synthesis.
c) Chronic consumption of NSAIDs results in nephrotoxicity that is reversible with time.
d) Amphotericin B nephrotoxicity can result in ADH-resistant polyuria.
e) Acetaminophen becomes nephrotoxic via activation by renal cytochrome P450.

Q.11. Which of the following statements is *not* correct?
a) Many nephrotoxicants appear to have the primary site of action on (in) the proximal tubule.
b) The proximal convoluted tubule is the primary site of reabsorption of glucose and amino acids.
c) The pars recta (S3) has a greater capacity to absorb organic compounds than the distal tubule.
d) The loop of Henle is the site of damage produced by chronic administration of analgesic mixtures.
e) The collecting duct appears relatively insensitive to most nephrotoxicants.

Q.12. Acetaminophen-induced liver injury is ________.
a) periportal
b) midzonal
c) centrolobular
d) biliary
e) diffuse

Q.13. The grayanotoxins found in many species of rhododendron produce effects similar to ________.
a) cyanogenic glycosides
b) veratrum alkaloids
c) lycopenes
d) disulfiram
e) taxol

Q.14. What is/are the most likely mechanism(s) by which glomerular nephrotoxicants such as antibiotics induce proteinurea?
a) By increasing the diameter of pores in the glomerular basement membrane
b) By enhancing carrier-mediated transport of protein across the glomerular membrane

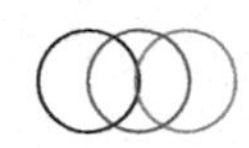

c) By inhibiting lysosomal protease activity
d) By reducing the number of fixed anionic charges on glomerular structural elements

Q.15. Which of the following is/are difference(s) in the chronic toxicity of inhaled elemental mercury and the toxicity of ingested mercurous salts?
a) Elemental mercury causes tremors, gingivitis, and increased excitability, whereas mercurous salts cause acrodynia, fever, and increased activity of sweat glands.
b) Mercurous salts produce renal toxicity similar to mercuric salts, whereas elemental mercury does not produce any renal toxicity.
c) Hypersensitivity reactions are more commonly associated with mercurous compounds than with mercury vapour.
d) Both compounds produce very similar chronic toxicity syndromes.

Q.16. Which of the following is *not* a characteristic of active transport?
a) Blocked by saxitoxin
b) Movement against a concentration gradient
c) Exhibits a transport maximum
d) Energy dependent
e) Selectivity

Q.17. Lyme disease is caused by which of the following?
a) *B. burgdorferi*
b) *H. capsulatum*
c) *M. tuberculosis*
d) *L. pneumophila*
e) *C. psittaci*

Q.18. Reabsorption of toxicants does *not* occur through ________.
a) entero-hepatic recirculation
b) glomerular filtration
c) diffusion
d) active transport
e) carriers for physiologic oxyanions

Q.19. Which of the following does *not* uncouple oxidative phosphorylation?
a) Pentachlorophenol
b) Dinitrophenol
c) Aconitase
d) Salicylate
e) Gramicidin

Q.20. Which of the following is *not* true regarding peroxisome proliferators?
a) They are a structurally diverse group of chemicals.
b) They cause a marked induction of lipid metabolizing enzymes.
c) They often are non-genotoxic hepatocarcinogens in rodents.
d) They induce hepatic CYP1A, which is indicative of peroxisome proliferation.
e) They operate via the peroxisome-proliferator-activated receptor.

Q.21. Which of the following is *least* likely to increase the occupational inhalation of a chemical?
a) Increased airborne concentration
b) Increased respiratory rate
c) Increased tidal volume
d) Increased particle size
e) Increased length of exposure

Q.22. Which would increase the likelihood of a toxic dosage through dermal exposure?
a) No preexisting skin disease.
b) Toxic exposure to thick skin
c) Increased percutaneous absorption rate

d) Low surface area of exposure
e) High epidermal intercellular junction integrity

Q.23. Which of the following lung diseases has the highest occupational death rate?
a) Asbestosis
b) Coal workers' pneumoconiosis
c) Byssinosis
d) Hypersensitivity pneumonitis
e) Silicosis

Q.24. Which of the following statements is FALSE regarding anaemia?
a) Alterations of the mean corpuscular volume are characteristic of anaemia.
b) Increased destruction of erythrocytes can lead to anaemia.
c) Decreased production of erythrocytes is not a common cause of anaemia because the bone marrow is continuously renewing the red blood cell pool.
d) Reticulocytes will live a longer period of time in the peripheral blood when a person is anaemic.
e) The main parameters in diagnosing anaemia are RBC count, haemoglobin concentration, and hematocrit.

Q.25. Which of the following types of anaemia is properly paired with its cause?
a) Iron deficiency anaemia–blood loss
b) Sideroblastic anaemia–vitamin B12 deficiency
c) Megaloblastic anaemia–folate supplementation
d) Aplastic anaemia–ethanol
e) Megaloblastic anaemia–lead poisoning

Q.26. The inability to synthesize the porphyrin ring of haemoglobin will most likely result in which of the following?
a) Iron-deficiency anaemia
b) Improper RBC mitosis
c) Inability to synthesize thymidine
d) Accumulation of iron within erythroblasts
e) Bone marrow hypoplasia

Q.27. Which of the following will cause a right shift in the oxygen dissociation curve?
a) Increased pH
b) Decreased carbon dioxide concentration
c) Decreased body temperature
d) Increased 2,3-BPG concentration
e) Fetal haemoglobin

Q.28. All of the following statements regarding erythrocytes are true *except* ________.
a) aged erythrocytes are removed by the liver, where the iron is recycled.
b) erythrocytes have a life span of approximately 120 days.
c) red blood cells generally lose their nuclei before entering the circulation.
d) reticulocytes are immature RBCs that still have a little RNA.
e) persons with anaemia have a higher than normal reticulocyte:erythrocyte ratio.

Q.29. All of the following statements regarding oxidative haemolysis are true *except* ________.
a) reactive oxygen species are commonly generated by RBC metabolism.
b) superoxide dismutase and catalase are enzymes that protect against oxidative damage.
c) reduced glutathione (GSH) increases the likelihood of oxidative injuries to RBCs.
d) glucose-6-phosphate dehydrogenase deficiency is commonly associated with oxidative haemolysis.
e) xenobiotics can cause oxidative injury to RBCs by overcoming the protective mechanisms of the cell.

Q.30. Which of the following sets of leukocytes is properly characterized as granulocytes because of the appearance of cytoplasmic granules on a blood smear?

 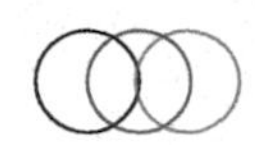

a) Neutrophils, basophils, and monocytes
b) Basophils, eosinophils, and lymphocytes
c) Eosinophils, neutrophils, and lymphocytes
d) Basophils, eosinophils, and neutrophils
e) Lymphocytes, neutrophils, and basophils

Answers (Exercise 1)

1. e	4. d	7. c	10. c	13. b	16. a	19. c	22. c	25. a	28. a
2. b	5. d	8. e	11. c	14. d	17. a	20. d	23. b	26. d	29. c
3. c	6. a	9. d	12. c	15. b	18. b	21. d	24. c	27. d	30. d

Exercise 2

Q.1. In which of the following locations would one *not* find melanin?
a) Iris
b) Ciliary body
c) Retinal pigment epithelium (RPE)
d) Uveal tract
e) Sclera

Q.2. Systemic exposure to drugs and chemicals is most likely to target which of the following retinal sites?
a) RPE and ganglion cell layer
b) Optic nerve and inner plexiform layer
c) RPE and photoreceptors
d) Photoreceptors and ganglion cell layer
e) Inner plexiform layer and RPE

Q.3. Which of the following structures is *not* part of the ocular fundus?
a) Retina
b) Lens
c) Choroid
d) Sclera
e) Optic nerve

Q.4. Drugs and chemicals in systemic blood have better access to which of the following sites because of the presence of loose endothelial junctions at that location?
a) Retinal choroid
b) Inner retina
c) Optic nerve
d) Iris
e) Ciliary body

Q.5. All of the following statements regarding ocular irritancy and toxicity are true *except* ________.
a) The Draize test involves instillation of a potentially toxic liquid or solid into the eye.
b) The effect of the irritant in the Draize test is scored on a weighted scale or the cornea, iris, and conjunctiva.
c) The Draize test usually uses one eye for testing and the other as a control.
d) The Draize test has a strong predictive value in humans.
e) The cornea is evaluated for opacity and area of involvement in the Draize test.

Q.6. Which of the following statements regarding colour vision deficits is FALSE?
a) Inheritance of a blue–yellow colour deficit is common.
b) Bilateral deficits in the visual cortex can lead to colour blindness.
c) Disorders of the outer retina produce blue–yellow deficits.
d) Drug and chemical exposure most commonly results in blue–yellow colour deficits.
e) Disorders of the optic nerve produce red–green deficits.

Q.7. A substance with which of the following pH values would be most damaging to the cornea?
  a) 1.0
  b) 3.0
  c) 7.0
  d) 10.0
  e) 12.0

Q.8. Which of the following statements concerning the lens is *false*?
  a) UV radiation exposure is a common environmental risk factor for developing cataracts.
  b) Cataracts are opacities of the lens that can occur at any age.
  c) The lens continues to grow throughout one's life.
  d) Naphthalene and organic solvents both can cause cataracts.
  e) Topical treatment with corticosteroids can cause cataracts.

Q.9. Which of the following is *not* a reason why the retina is highly vulnerable to toxicant-induced damage?
  a) Presence of numerous neurotransmitter systems
  b) Presence of melanin in the RPE
  c) High choroidal blood flow rate
  d) High rate of oxidative mitochondrial metabolism
  e) Lack of gap junctions

Q.10. A deficiency in which of the following vitamins can result in degeneration of optic nerve fibres?
  a) Vitamin A
  b) Vitamin B3
  c) Vitamin C
  d) Vitamin B12
  e) Vitamin E

Q.11. Which of the following statements regarding axons and/or axonal transport is *false*?
  a) Single nerve cells can be over 1 m in length.
  b) Fast axonal transport is responsible or movement of proteins from the cell body to the axon.
  c) Anterograde transport is accomplished by the protein kinesin.
  d) The motor proteins, kinesin a dynein, are associate with microtubules.
  e) A majority of the ATP in nerve cells is used for axonal transport.

Q.12. Which of the following statements is not characteristic of Schwann cells in Wallerian degeneration?
  a) Schwann cells provide physical guidance needed for the regrowth of the axon.
  b) Schwann cells release trophic factors that stimulate growth.
  c) Schwann cells act to clear the myelin debris with the help of macrophages.
  d) Schwann cells increase synthesis of myelin lipids in response to axonal damage.
  e) Schwann cells are responsible for myelination of axons in the peripheral nervous system.

Q.13. Prenatal exposure to ethanol can result in mental retardation and hearing deficits in the newborn. What is the cellular basis of the neurotoxicity?
  a) Neuronal loss in cerebellum
  b) Acute cortical haemorrhage
  c) Microcephaly
  d) Loss of hippocampal neurons
  e) Degeneration of the basal ganglia

Q.14. Which of the following characteristics is *least* likely to place a neuron at risk of toxic damage?
  a) High metabolic rate
  b) Ability to release neurotransmitters

 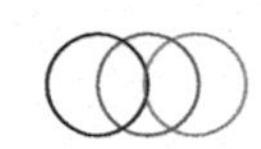

c) Long neuronal processes supported by the soma
d) Excitable membranes
e) Large surface area

Q.15. The use of meperidine contaminate with MPTP will result in a Parkinson's disease-like neurotoxicity. Where is the most likely site in the brain that MPTP exerts its toxic effects?
a) Cerebellum
b) Cerebral cortex
c) Brainstem
d) Substantia nigra
e) Hippocampus

Q.16. Which of the following statements regarding the PNS and the CNS is correct?
a) Nerve-impulse transduction is much faster in the CNS than in the PNS.
b) PNS axons can regenerate, whereas CNS axons cannot.
c) Remyelination does not occur in the CNS.
d) Oligodendrocytes perform remyelination in the PNS.
e) In the CNS, oligodendrocyte scarring interferes with axonal regeneration.

Q.17. Platinum (cisplatin) results in which of the following neurologic problems?
a) Peripheral neuropathy
b) Trigeminal neuralgia
c) Spasticity
d) Gait ataxia
e) Tremor

Q.18. Which of the following is *not* characteristic of axonopathies?
a) There is degeneration of the axon.
b) The cell body of the neuron remains intact.
c) Axonopathies result from chemical transaction of the axon.
d) A majority of axonal toxicants caused motor deficits.
e) Sensory and motor deficits are first noticed in the hands and feet following axonal degeneration.

Q.19. All of the following statements regarding lead exposure are true *except* ________.
a) Lead exposure results in peripheral neuropathy.
b) Lead slows peripheral nerve conduction in humans.
c) Lead causes the transection of peripheral axons.
d) Segmental demyelination is a common result of lead ingestion.
e) Lead toxicity can result in anaemia.

Q.20. Regarding excitatory amino acids, which of the following statements is *false*?
a) Glutamate is the most common excitatory amino acid in the CNS.
b) Excitotoxicity has been linked to conditions such as epilepsy.
c) Overconsumption of monosodium glutamate (MSG) can result in a tingling or burning sensation in the face and neck.
d) An ionotropic glutamate receptor is coupled to a G protein.
e) Glutamate is toxic to neurons.

Q.21. Which of the following cell types secretes anti-Müllerian hormone (AMH)?
a) Spermatogonium
b) Leydig cell
c) Sertoli cell
d) Primary spermatocyte
e) Spermatid

Q.22. Penile erections are dependent on ________.
a) the CNS
b) sympathetic nerve stimulation
c) helicine (penile) artery constriction

d) corpora cavernosum smooth muscle relaxation
e) a spinal reflex arc

Q.23. The corpus luteum is responsible for the secretion of which of the following hormones during the first part of pregnancy?
a) Estradiol and hCG
b) Progesterone and estradiol
c) Progesterone and hCG
d) FSH and LH
e) FSH and progesterone

Q.24. All of the following statements regarding the hypothalamopituitary–gonadal axis are true *except* ________.
a) FSH increases testosterone production by the Leydig cells.
b) FSH and LH are synthesized in the anterior pituitary.
c) Estradiol provides negative feedback on the hypothalamus and the anterior pituitary.
d) GnRH from the hypothalamus increases FSH and LH release from the anterior pituitary.
e) The LH spike during the menstrual cycle is responsible for ovulation.

Q.25. Which of the following statements is FALSE regarding gametal DNA repair?
a) DNA repair in spermatogenic cells is dependent on the dose of chemical.
b) Spermatogenic cells are less able to repair damage from alkylating agents.
c) Female gametes have base excision repair capacity.
d) Meiotic maturation of the oocyte decreases its ability to repair DNA damage.
e) Mature oocytes and mature sperm no longer have the ability to repair DNA damage.

Q.26. Reduction division takes place during the transition between which two cell types during spermatogenesis?
a) Spermatogonium and primary spermatocyte
b) Primary spermatocyte and secondary spermatocyte
c) Secondary spermatocyte and spermatid
d) Spermatid and spermatozoon
e) Spermatozoon and mature sperm

Q.27. Which of the following cell types is properly paired with the substance that it secretes?
a) Ovarian granulosa cells–progesterone
b) Leydig cells–ABP
c) Ovarian thecal cells–estrogens
d) Sertoli cells–testosterone
e) Gonadotrophin–LH

Q.28. Which of the following statements regarding male reproductive capacity is *false*?
a) Kline Elder's syndrome males are sterile.
b) FSH levels are often measured in order to determine male reproductive toxicity of a particular toxin.
c) Divalent metal ions, such as Hg and Cu, act as androgen receptor antagonists and affect male reproduction.
d) The number of sperms produced per day is approximately the same in all males.
e) ABP is an important biochemical marker for testicular injury.

Q.29. Reduction of sperm production can be caused by all of the following diseases *except* ________.
a) Hypothyroidism
b) Measles
c) Crohn's disease
d) Renal failure
e) Mumps

Q.30. Of the following, which is *least* likely to be affected by estrogen?
a) Nervous system
b) Musculoskeletal system

 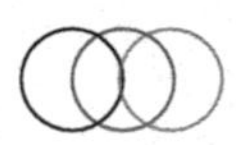

c) Digestive system
d) Cardiovascular system
e) Urinary system

Answers (Exercise 2)

1. e	4. a	7. e	10. d	13. c	16. b	19. c	22. d	25. e	28. d
2. c	5. d	8. d	11. e	14. b	17. a	20. d	23. b	26. b	29. b
3. b	6. a	9. e	12. d	15. d	18. d	21. c	24. a	27. e	30. c

Exercise 3

Q.1. In which of the following locations would one *not* find spontaneous depolarization?
a) SA node
b) Myocardium
c) AV node
d) Bundle of His
e) Purkinje fibres

Q.2. Which of the following scenarios would increase contractility of the myocardium?
a) Increased activity of the Na^+/K^+ -ATPase
b) Increased activity of sacroplasmic reticulum Ca^{2+} ATPase
c) Decreased activity of sacroplasmic reticulum Ca^{2+} ATPase
d) Decreased intracellular calcium levels
e) Increased intracellular K^+ levels

Q.3. All of the following statements regarding abnormal cardiac function are true *except* ________.
a) Ventricular arrhythmias are generally more severe than atrial arrhythmias.
b) Ventricular hypertrophy is a common cause of ventricular arrhythmias.
c) Coronary artery atherosclerosis is a major cause of ischemic heart disease.
d) Right-sided heart failure results in pulmonary edema.
e) Tachycardia is classified as a rapid resting heart rate (>100 beats/min).

Q.4. Ion balance is very important in maintaining a normal cardiac rhythm. Which of the following statements is ________?
a) Blockade of K^+ channels decreases the duration of the action potential.
b) Blockade of Ca^{2+} channels has a positive inotropic effect.
c) Inhibition of Na^+ channels increases conduction velocity.
d) Blockage of the Na^+/K^+-ATPase increases contractility.
e) Calcium is transported into the cell via a Ca^{2+}-ATPase.

Q.5. Which of the following is most likely *not* a cause of myocardial reperfusion injury?
a) Cellular pH fluctuations
b) Damage to the sarcolemma
c) Generation of toxic oxygen radicals
d) Ca^{2+} overload
e) Inhibition of the electron transport chain

Q.6. Which of the following statements regarding the cardiotoxic manifestations of ethanol consumption is *false*?
a) Acute ethanol toxicity causes decreased conductivity.
b) Chronic alcohol consumption is associated with arrhythmias.
c) Acute ethanol toxicity causes an increased threshold for ventricular fibrillation.
d) Chronic ethanol toxicity can result in cardiomyopathy.
e) Acetaldehyde is a mediator of cardiotoxicity.

Q.7. Cardiac glycosides ________.
a) increase the activity of the Na^+/K^+-ATPase
b) make the resting membrane potential more negative

c) can have sympathomimetic and parasympathomimetic effects
d) decrease ventricular contractility
e) increase AV conduction

Q.8. Which of the following is *not* a common cardiotoxic manifestation of cocaine abuse?
a) Parasympathomimetic effects
b) Myocardial infarction
c) Cardiac myocyte death
d) Ventricular fibrillation
e) Ischemia

Q.9. Using high doses of anabolic–androgenic steroids is *not* likely associated with which of the following?
a) An increase in LDL
b) Cardiac hypertrophy
c) Myocardial infarction
d) Increased nitric oxide synthase expression
e) A decrease in HDL

Q.10. Which of the following is *not* a common mechanism of vascular toxicity?
a) Membrane disruption
b) Oxidative stress
c) Bioactivation of protoxicants
d) Reduction and accumulation of LDL in endothelium
e) Accumulation of toxin in vascular cells

Q.11. Which of the following cells or substances is *not* part of the innate immune system?
a) Lysozyme
b) Monocytes
c) Complement
d) Antibodies
e) Neutrophils

Q.12. Myeloid precursor stem cells are responsible or the formation of all of the following *except* ________.
a) platelets
b) lymphocytes
c) basophils
d) erythrocytes
e) monocytes

Q.13. When an Rh^- mother is exposed to the blood of a Rh^+ baby during childbirth, the mother will make antibodies against the Rh factor, which can lead to the mother attacking the next Rh^+ fetus. This is possible because of which antibody's ability to cross the placenta?
a) IgM
b) IgE
c) IgG
d) IgA
e) IgD

Q.14. Which of the following statements is *false* regarding important cytokine function in regulating the immune system?
a) IL-1 induces inflammation and fever.
b) IL-3 is the primary T-cell growth factor.
c) IL-4 induces B-cell differentiation and isotype switching.
d) Transforming growth factor-β (TGF-β) enhances monocyte/macrophage chemotaxis.
e) Interferon gamma (IFN-gamma) activates macrophages.

 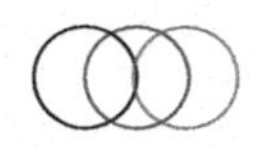

Q.15. Which of the following is *not* a step performed during an enzyme-linked immunosorbent assay (ELISA)?
  a) A chromogen is added and colour is detected.
  b) The antigen of interest is fixed to a micro titer plate.
  c) Radioactively labelled cells are added to the solution.
  d) Enzyme-tagged secondary antibodies are added.
  e) Test sera are added.

Q.16. The delayed hypersensitivity response (DHR) test does *not*________.
  a) evaluate memory T-cells' ability to recognize a foreign antigen
  b) evaluate memory T-cells' ability to secrete cytokines
  c) evaluate memory T-cells' ability to proliferate
  d) evaluate memory T-cells' ability to lyze foreign target cells
  e) evaluate memory T-cells' ability to migrate to the site of foreign antigen

Q.17. The number of alveolar macrophages in smokers is greatly increased relative to non-smokers. What is a characteristic of the alveolar macrophages found in smokers?
  a) They are in an inactive state.
  b) They are far larger than normal.
  c) They have increased phagocytic activity.
  d) They are incapable of producing cytokines.
  e) They have decreased bactericidal activity.

Q.18. Which of the following is *not* characteristic of a type I hypersensitivity reaction?
  a) It is mediated by IgE.
  b) It involves immune complex deposition in peripheral tissues.
  c) It involves mast-cell degranulation.
  d) Anaphylaxis is an acute, systemic, and very severe type I hypersensitivity reaction.
  e) It is usually mediated by preformed histamine, prostaglandins, and leukotrienes.

Q.19. Which of the following types of hypersensitivity is *not* mediated by antibodies?
  a) Type I
  b) Type II
  c) Type III
  d) Type IV
  e) Type V

Q.20. Which of the following is *not* a common mechanism of autoimmune disorders?
  a) Subjection to positive selection in the thymus
  b) Anergic T cells become activated
  c) Interference with normal immunoregulation by $CD8^+$ suppressor cells
  d) Lack of subjection to negative selection in the thymus
  e) Decreased self-tolerance

Q.21. Which of the following statements is FALSE regarding the role of mucus in the conducting airways?
  a) Pollutants trapped by mucus can be eliminated via expectoration or swallowing.
  b) Mucus is of a basic pH.
  c) The beating of cilia propels mucus out of the lungs.
  d) Mucus plays a role promoting oxidative stress.
  e) Free radical scavenging is believed to be a role of mucus.

Q.22. Respiratory distress syndrome sometimes affects premature neonates due to lack of surfactant production by which of the following cell types?
  a) Lung fibroblasts
  b) Type II pneumocytes
  c) Endothelial cells

d) Alveolar macrophages
e) Type I pneumocytes

Q.23. In a situation where there is an increased metabolic demand for oxygen, which of the following volume measurements will greatly increase?
a) Total lung capacity (TLC)
b) Residual volume (RV)
c) Functional residual capacity (FRC)
d) Tidal volume (TV)
e) Vital capacity (VC)

Q.24. The free radicals that inflict oxidative damage on the lungs are generated by all of the following *except* ________.
a) tobacco smoke
b) neutrophils
c) ozone
d) monocytes
e) SO_2

Q.25. Which of the following gases would most likely pass all the way through the respiratory tract and diffuse into the pulmonary blood supply?
a) O_3 (ozone)
b) NO_2
c) H_2O
d) CO
e) SO_2

Q.26. All of the following statements regarding particle deposition and clearance are true *except*
a) One of the main modes of particle clearance is via mucociliary escalation.
b) Diffusion is important in the deposition of particles in the bronchial regions.
c) Larger volumes of inspired air increase particle deposition in the airways.
d) Sedimentation results in deposition in the bronchioles.
e) Swallowing is an important mechanism of particle clearance.

Q.27. Which of the following is not a common location to which particles are cleared?
a) Stomach
b) Lymph nodes
c) Pulmonary vasculature
d) Liver
e) GI tract

Q.28. Pulmonary fibrosis is marked by which of the following?
a) Increased type I collagen
b) Decreased type III collagen
c) Increased compliance
d) Elastase activation
e) Decreased overall collagen levels

Q.29. Activation of what enzyme(s) is responsible or emphysema?
a) Antitrypsin
b) Epoxide hydrolase
c) Elastase
d) Hyaluronidase
e) Nonspecific proteases

Q.30. Which of the following measurements would *not* be expected from a patient with restrictive lung disease?
a) Decreased FRC
b) Decreased RV

 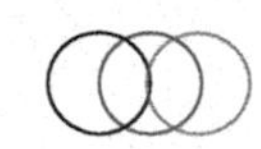

c) Increased VC
d) Decreased FEV1
e) Impaired ventilation

Answers (Exercise 3)

1. b	4. d	7. c	10. d	13. c	16. d	19. d	22. b	25. d	28. a
2. b	5. e	8. a	11. d	14. b	17. e	20. a	23. d	26. b	29. c
3. d	6. c	9. d	12. b	15. c	18. b	21. d	24. e	27. d	30. c

Exercise 4

Q.1. The FDA protocol that primarily examines fertility and preimplantation and postimplantation viability is ________.
a) Segment I
b) Segment II
c) Segment III
d) Segment IV

Q.2. The FDA protocol that primarily examines postnatal, survival, growth, and external morphology is ________.
a) Segment I
b) Segment II
c) Segment III
d) Segment IV

Q.3. All of the following maternal diseases have been assumed with adverse pregnancy outcomes *except* ________.
a) allergic rhinitis
b) febrile illness during the first trimester
c) hypertension
d) diabetes mellitus

Q.4. All of the following are true of the development toxicity of cadmium *except* ________.
a) It appears to involve placental toxicity.
b) It appears to involve inhibition of nutrient transport across the placenta.
c) Zinc can affect the developmental toxicity of cadmium.
d) Cadmium induces transferrin, which binds zinc in the placenta.

Q.5. All of the following are necessary for a normally developing embryo *except* ________.
a) apoptosis
b) cell proliferations
c) cell differentiation
d) necrosis

Q.6. The fetal period is characterized by all of the following *except* ________.
a) beginning organ development
b) tissue differentiation
c) growth
d) physiological maturation

Q.7. Toxic exposure during the fetal period is likely to affect ________.
a) organogenesis
b) implantation
c) growth and maturation
d) all of the above

Q.8. Approximately what percentage of marked drugs belong to the FDA pregnancy category?
 a) 1%
 b) 10%
 c) 20%
 d) 30%

Q.9. Leukaemias ________.
 a) are often due to cytogenic abnormalities, particularly damage to or loss of chromosomes 8 and 11
 b) are rarely caused by agents used in cancer chemotherapy
 c) originate in circulating blood cells
 d) are characterized as 'acute' if their effects are short-lived and severe
 e) have long been associated with exposure to x-ray radiation

Q.10. Regarding platelets and thrombocytopenia, which of the following statements is *false*?
 a) Platelets can be removed from the circulation through a hapten-mediated pathway that is induced by drugs or chemicals.
 b) Cortisol decreases platelet activity by inhibiting thromboxane prostaglandin synthesis.
 c) Toxins can induce a change in a platelet membrane glycoprotein, leading to recognition and removal of the platelet by phagocytes.
 d) Heparin administration can result in platelet aggregation and cause thrombocytopenia.
 e) Thrombotic thrombocytopenic purpura is most commonly caused by infectious disease but can also be associated with administration of pharmacologic agents.

Q.11. Which of the following statements is *false* regarding skin histology?
 a) Blood supply to the epidermis originates in the epidermal–dermal junction.
 b) Melanin is made and stored by melanocytes.
 c) The stratum corneum is made up of nonviable cells.
 d) It takes approximately 2 weeks for cells to be sloughed off from the stratum corneum.
 e) Stem cells in the basal layer replenish the keratinocytes of the layers of epidermis.

Q.12. Transdermal drug delivery does *not*________.
 a) prevent drug exposure to low pH
 b) avoid first-pass metabolism
 c) provide steady infusion over an extended period of time
 d) avoid large variation in drug plasma concentration
 e) increase safety of drug delivery

Q.13. Irritant and contact dermatitis are marked by all of the following characteristics *except* ________.
 a) softness
 b) erythema
 c) flaking
 d) induration
 e) blistering

Q.14. Nickel is a common cause of allergic contact dermatitis, which is which type of hypersensitivity reaction?
 a) Type I
 b) Type II
 c) Type III
 d) Type IV
 e) Type V

Q.15. All of the following statements regarding phototoxicology are true *except* ________.
 a) Melanin is primarily responsible or the absorption of UV-B radiation.
 b) UV-A is the most effective at causing sunburn in humans.
 c) IL-1 release is responsible or systemic symptoms associated with sunburn.

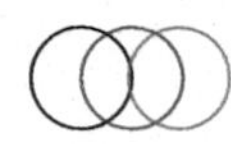

d) Melanin darkening is a common response to UV exposure.
e) UV radiation exposure causes thickening of the stratum corneum.

Q.16. Photo allergies ________.
a) represent a form of type III hypersensitivity reaction
b) can occur without exposure to UV radiation
c) are hapten-mediated
d) cannot be tested for as contact dermatitis allergies can
e) often occur on first exposure

Q.17. Diffusion through the epidermis would occur most slowly across skin at which of the following locations?
a) Palm
b) Forehead
c) Scrotum
d) Foot sole
e) Abdomen

Q.18. Which of the following statements regarding photosensitivity is *false*?
a) Porphyria cause light sensitivity because of the lack of heme synthesis.
b) Lupus patients are unable to repair damage caused by UV light.
c) Chronic phototoxic responses often result in hyperpigmentation.
d) Photo allergy represents a type IV hypersensitivity reaction.
e) UV radiation causes cycloadducts between pyrimidine bases.

Q.19. Acne is caused by all of the following *except* ________.
a) clogged sebaceous glands
b) hormones
c) viruses
d) genetics
e) environmental actors

Q.20. All of the following statements regarding urticaria are true *except* ________.
a) Urticaria is a delayed-type hypersensitivity reaction.
b) Hives are mediated partly by histamine release from mast cells.
c) Latex is a common chemical cause of urticaria.
d) Select foods have been reported to elicit contact urticaria.
e) Urticaria is mediated by IgE antibodies.

Q.21. Which of the following is thought to be an important factor in the pathology of alcohol-induced liver disease?
a) Inflammatory response
b) Lipid peroxidation
c) Oxidative stress
d) All of the above

Q.22. Allyl alcohol is metabolized by ADH to ________.
a) Benzylaldehyde
b) Acrolein
c) Acetic anhydride
d) Butyraldehyde

Q.23. All of the following statements are true *except* ________.
a) Xenobiotics can greatly slow down the proliferation of neutrophils and monocytes, increasing the risk of infection.
b) Ethanol and cortisol decrease phagocytosis and microbe ingestion by the immune system.
c) Agranulocytosis is predictable and can be caused by exposure to a number of environmental toxins.

d) Heroin and methadone abusers have reduced the ability to kill microorganisms due to a drug-induced reduction in superoxide production.
e) Toxic neutropenia may be mediated by the immune system.

Q.24. Ethyl alcohol is metabolized in humans by all of the following *except* ________.
a) CYP3A4
b) CYP2E1
c) ADH
d) peroxisome catalase

Q.25. Idiosyncratic liver injury is characterized by all of the following *except* ________.
a) can be immune or non-immune mediated
b) has a clear dose-response relationship
c) is relatively rare
d) has a probable genetic basis

Q.26. The liver-cell process associated with cell swelling leakage of cell contents, and an influx of inflammatory cells is ________.
a) apoptosis
b) fibrosis
c) necrosis
d) steatosis

Q.27. All of the following are true regarding the hepatotoxicity of carbon tetrachloride *except* ________.
a) The reactive metabolite is formed by cytochrome P450 3A4.
b) The reactive metabolite is a free radical.
c) Chronic ethanol exposure can enhance the injury.
d) The injury involves lipid peroxidation.

Q.28. All of the following are true regarding ethanol and the liver *except* ________.
a) Ethanol inhibits the transfer of triglycerides from liver to adipose tissue.
b) Alcohol dehydrogenase is the only inducible enzyme in chronic alcoholism.
c) An inactive form of aldehydrogenase is formed in 50% of Asians.
d) The catalase pathway is a minor route for ethanol metabolism.

Q.29. All of the following hepatic sites are matched with the appropriate preferential toxicant *except* ________.
a) zone1 hepatocyte–iron
b) bile duct cells–ethanol
c) stellate cells–vitamin A
d) zone 3 hepatocyte–carbon tetrachloride

Q.30. All of the following cause non-immune idiocyncratic live toxicity *except* ________.
a) tienilic acid
b) isoniazid
c) amiodarone
d) ketoconazole

Answers (Exercise 4)

1. a	4. d	7. c	10. b	13. a	16. c	19. c	22. b	25. b	28. b
2. c	5. d	8. a	11. b	14. d	17. d	20. a	23. c	26. c	29. b
3. a	6. a	9. e	12. e	15. b	18. c	21. d	24. a	27. a	30. a

 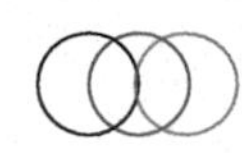

Exercise 5

Q.1. Which of the following is associated with pathological hypertrophy of the heart?
- a) Hypertension
- b) Exercise
- c) Pregnancy
- d) None of the above

Q.2. Myocardial accumulation of collagen is not associated with ________.
- a) ischemic cardiomyopathy
- b) myocardial infarction
- c) pathological hypertrophy
- d) adaptive hypertrophy

Q.3. Counterregulatory mechanism in response to compensatory mechanisms to cardiac hypertrophy leads to ________.
- a) decrease in heart size
- b) myocardial remodelling
- c) decrease in cardiac fibrosis
- d) decrease in salt/water retention

Q.4. Inflammatory lesions in the vascular system are termed ________.
- a) vasculitis
- b) embolitis
- c) thrombitis
- d) angio-inflammation

Q.5. The most prevalent vascular structural injury is ________.
- a) capillary hyperplasia
- b) varicose veins
- c) angioma
- d) atherosclerosis

Q.6. Cardiac glycosides like digoxin ________.
- a) increase the sensitivity of myocytes to calcium
- b) inhibit sodium/potassium ATPase
- c) causes sinus tachycardia
- d) inhibit sympathetic outflow at high doses

Q.7. Which of the following is the most sensitive clinical indicator of myocardial cell damage?
- a) Urine CPK
- b) Serum troponin
- c) Serum ALT
- d) Serum creatinine

Q.8. Which of the following is an indicator of fluid overload in congestive heart failure?
- a) Urine pH
- b) Serum CPK
- c) Brain natriuretic peptide (BNP)
- d) First degree AV block on ECG

Q.9. Moxifloxacin is associated with ________.
- a) acute congestive heart failure
- b) prolongation of the QT interval
- c) coronary artery thrombotic events
- d) toxic cardiomyopathy

Q.10. COX-2 inhibits, presumably, and increases the risk for cardiovascular events by causing ________.
- a) heart block
- b) systolic disfunction

c) toxic cardiomyopathy
d) coronary artery thrombotic events

Q.11. The impairment of hepatic function can have numerous negative consequences. Which of the following is likely *not* caused by impaired hepatic function?
a) Jaundice
b) Hypercholesterolemia
c) Hyperammonemia
d) Hyperglycemia
e) Hypoalbuminemia

Q.12. All of the following statements regarding the liver are true *except* ________.
a) The major role of the liver is to maintain metabolic homeostasis of the body.
b) The liver encounters ingested nutrients before the heart does.
c) Hepatic triads contain a branch of the hepatic portal vein, a branch of the hepatic artery, and a bile ductule.
d) The liver manufactures and stores bile.
e) The large fenestrae of hepatic sinusoids facilitate the exchange of materials between the sinusoid and the hepatocyte.

Q.13. Activation of which of the following cell types can result in increased secretion of collagen scar tissue, leading to cirrhosis?
a) Hepatocyte
b) Ito cell
c) Kupffer cell
d) Endothelial cell
e) β-cell

Q.14. Wilson's disease is a rare genetic disorder characterized by the failure to export which of the following metals into bile?
a) Iron
b) Zinc
c) Silver
d) Lead
e) Copper

Q.15. Which of the following is *not* characteristic of apoptosis?
a) Cell swelling
b) Nuclear fragmentation
c) Lack of inflammation
d) Programmed death
e) Chromatin condensation

Q.16. A patient suffering from canalicular cholestasis would *not* be expected to exhibit which of the following?
a) Increased bile salt serum levels
b) Jaundice
c) Increased bile formation
d) Dark brown urine
e) Vitamin A deficiency

Q.17. Which of the following statements regarding liver injury is *false*?
a) Large doses of acetaminophen have been shown to cause a blockade of hepatic sinusoids.
b) Hydrophilic drugs readily diffuse into hepatocytes because of the large sinusoidal fenestrations.
c) There are sinusoidal transporters that take toxicants up into hepatocytes.
d) Hepatocellular cancer has been associated with androgen abuse.
e) In cirrhosis, excess collagen is laid down in response to direct injury or inflammation.

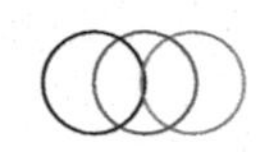

Q.18. The inheritance of a 'slow' aldehyde dehydrogenase enzyme would result in which of the following after the ingestion of ethanol?
a) High ethanol tolerance
b) Little response to low doses of ethanol
c) Low serum levels of acetaldehyde
d) Nausea
e) Increased levels of blood ethanol compared to an individual with a normal aldehyde dehydrogenase

Q.19. Which of the following is not a common mechanism of hepatocellular injury?
a) Deformation of the hepatocyte cytoskeleton
b) Mitochondrial injury
c) Cholestasis
d) Interference with vesicular transport
e) Increased transcytosis between hepatocytes

Q.20. Ethanol is not known to cause which of the following types of hepatobiliary injury?
a) Fatty liver
b) Hepatocyte death
c) Fibrosis
d) Immune-mediated responses
e) Canalicular cholestasis

Q.21. The percentage of mating resulting in pregnancy is called ________.
a) fertility index
b) gestation index
c) viability index
d) survival index

Q.22. The percentage of pregnancies resulting in live litters is ________.
a) fertility index
b) gestation index
c) viability index
d) survival index

Q.23. The lactation index in rats is the ———.
a) number of live births that breastfeed
b) number of days an animal breastfeeds
c) calories lost per day by a mother who breastfeeds
d) percentage of animals alive at 4 days that survive the 21-day lactation period

Q.24. Which of the following statement is *false*?
a) There is good concordance between human and animal neurotoxicity assessment.
b) The developing nervous system is insensitive to toxicant exposure.
c) Monkeys can be used to test low-level effects of neurotoxic cells.
d) In vitro b cell cultures can be used in neurotoxicity evaluation.

Q.25. A severe cytokine response that progressed in systemic organ failure occurred in a Phase 1 study involving the use of ________.
a) an uncoupler of oxidative phosphorylation
b) a COX-2 inhibitor
c) a CD 28 monoclonal antibody
d) a microtubule assembly inhibitor

Q.26. It has been postulated that within the human genome, how much variability in the DNA sequence exists between any two individuals?
a) 0.01%
b) 0.1%
c) 0.5%
d) 1.0%

Q.27. Methyl bromide (CH_3Br) ________.
a) is a liquid used primarily as a fumigant
b) has essentially no warning properties, even at physiologically hazardous concentrations
c) is extremely flammable
d) is of greater concern from its oral toxicity than from its inhalation toxicity
e) would not be expected to be readily absorbed through the lungs

Q.28. The chloronicotinyl compound imidacloprid demonstrates a high insecticidal potency and exceptionally low mammalian toxicity due to ________.
a) its high affinity for insect nicotinic acetylcholine receptors and low affinity for mammalian nicotinic acetylcholine receptors
b) the blood–brain barrier in mammals
c) the first-pass effect in the liver in mammals
d) the low pH in the stomach of monogastric mammals
e) the presence of acetylcholinesterase in mammals

Q.29. The most commonly used pyrethroid synergist is ________.
a) silica
b) piperonyl butoxide
c) methyl butyl ether
d) *n*-octyl bicyloheptene dicarboximide
e) toluene

Q.30. Paraquat and diquat differ substantially in their ________.
a) metabolism to a free radical
b) ability to initiate lipid peroxidation in vivo
c) uptake by the lung
d) generation of superoxide anion in vivo
e) mechanism of cytotoxicity

Answers (Exercise 5)

1. a	4. a	7. b	10. d	13. b	16. c	19. e	22. b	25. c	28. a
2. d	5. d	8. c	11. d	14. e	17. b	20. e	23. d	26. b	29. b
3. b	6. b	9. b	12. d	15. a	18. d	21. a	24. b	27. b	30. c

Exercise 6

Q.1. Which of the following assumptions is *not* correct regarding risk assessment for male reproductive effects in the absence of mechanistic data?
a) An agent that produces an adverse reproductive effect in experimental animals is assumed to pose a potential reproductive hazard to humans.
b) In general, a non-threshold is assumed for the dose-response curve for male reproductive toxicity.
c) Effects of xenobiotics on male reproduction are assumed to be similar across species unless demonstrated otherwise.
d) The most sensitive species should be used to estimate human risk.
e) Reproductive processes are similar across mammalian species.

Q.2. The most serious consequence of crude oil or kerosene ingestion by cattle is ________.
a) liver damage
b) kidney damage
c) aspiration pneumonia
d) central nervous system stimulation
e) leukaemia

Q.3. Toxic injury to the cell body, axon, and surrounding Schwann cells of peripheral nerves are referred to, respectively, as ________.
a) neuropathy, axonopathy, and myelopathy
b) neuronopathy, axonopathy, and myelinopathy

 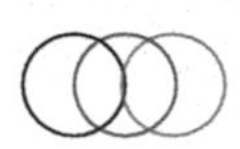

c) neuropathy, axonopathy, and gliosis
d) neuronopathy, dying-back neuropathy, and myelopathy
e) chromatolysis, axonopathy, and gliosis

Q.4. The conceptus is considered preferentially vulnerable to chemical insult due to all of the following *except* ________.
a) rapid rate of cell proliferation
b) requirement for precise temporal and spatial localization of cells and cell products
c) rapid blood flow (heart rate) and increased tissue distribution of the chemical
d) limited drug metabolism capability
e) immaturity of the immune system

Q.5. Which of the following is a non-genotoxic liver carcinogen in rats?
a) Aflatoxin
b) Vinyl chloride
c) Pyrrolizidine alkaloids
d) Clofibrate
e) Tamoxifen

Q.6. Suppression of NK (natural killer) cell activity produces adverse effects in animals because ________.
a) NK cells play a role in the immune surveillance of tumours and the inhibition of metastases.
b) NK cells are precursors to pulmonary macrophages, and suppression of activity results in a decreased ability to combat pulmonary infections.
c) NK cells function as a helper cell in erythropoiesis, and suppression of NK cell activity results in anaemia.
d) NK cells function in the complement-fixation cascade, and reduced NK cell activity causes loss of blood clotting.
e) NK cells function directly by phagocytosis of pathogens to protect the host from pathogenic bacteria (acquired cell-mediated immunity).

Q.7. Which of the following is *not* about arsine?
a) It is a gas at room temperature.
b) It produces acute intravascular haemolysis.
c) It has a garlic-like odor.
d) Acute renal failure is a common manifestation of arsine poisoning.
e) Significant hepatotoxicity often occurs as part of arsine poisoning.

Q.8. No specific antidote is available for poisoning by ________.
a) sodium fluoroacetate
b) warfarin
c) chlorinated hydrocarbon insecticides
d) rotenone
e) cyanide

Q.9. Which of the following is *not* commonly associated with mercury vapour poisoning?
a) Acute, corrosive bronchitis
b) Interstitial pneumonitis
c) Tremor
d) Increased excitability
e) Vomiting and bloody diarrhoea

Q.10. Compounds containing beryllium are *not* readily absorbed from the gastrointestinal tract because ________.
a) they form insoluble phosphate precipitates at the pH of the intestinal tract
b) they are chelated by bile salts in the small intestine

c) their size prevents passage through cell membranes
d) they are converted to an oxide form
e) exposures rarely occur to the soluble metallic form

Q.11. The inability to release hormones from the anterior pituitary would *not* affect the release of which of the following?
a) LH
b) PRL
c) ADH
d) SH
e) ACTH

Q.12. Which of the following statements regarding pituitary hormones is ________?
a) The hypothalamic–hypophyseal portal system transports releasing hormones to the neurohypophysis.
b) Dopamine enhances prolactin secretion from the anterior pituitary.
c) Somatostatin inhibits the release of GH.
d) The function of chromophobes in the anterior pituitary is unknown.
e) Oxytocin and ADH are synthesized by hypothalamic nuclei.

Q.13. 21-Hydroxylase deficiency causes masculinization of female genitals at birth by increasing androgen secretion from which region of the adrenal gland?
a) Zona glomerulosa
b) Zona reticularis
c) Adrenal medulla
d) Zona asciculata
e) Chromaffin cells

Q.14. Which of the following statements regarding adrenal toxicity is ________?
a) The adrenal cortex and adrenal medulla are equally susceptible to at-soluble toxins.
b) Adrenal cortical cells lack the enzymes necessary to metabolize xenobiotic chemicals.
c) Pheochromocytoma of the adrenal medulla can cause high blood pressure and clammy skin due to increased epinephrine release.
d) Xenobiotics primarily affect the hydroxylase enzymes in the zona reticularis.
e) Vitamin D is an important stimulus for adrenal cortex steroid secretion.

Q.15. Chemical blockage of iodine transport in the thyroid gland ________.
a) affects export of T3 and T4
b) prevents reduction to I2 by thyroid peroxidase
c) decreases RH release from the hypothalamus
d) interrupts intracellular thyroid biosynthesis
e) mimics goiter

Q.16. Chromaffin cells of the adrenal gland are responsible for secretion of which of the following?
a) Aldosterone
b) Epinephrine
c) Corticosterone
d) Testosterone
e) Estradiol

Q.17. The para follicular cells of the thyroid gland are responsible for secreting a hormone that ________.
a) increases blood glucose levels
b) decreases plasma sodium levels
c) increases calcium storage
d) decreases metabolic rate
e) increases bone resorption

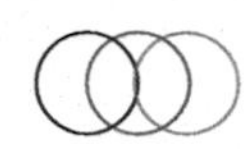

Q.18. Parathyroid adenomas resulting in increased pH levels would be expected to cause which of the following?
a) Hypocalcemia
b) Hyperphosphatemia
c) Increased bone formation
d) Osteoporosis
e) Rickets

Q.19. Which of the following vitamins increases calcium and phosphorus absorption in the gut?
a) Vitamin D
b) Niacin
c) Vitamin A
d) Vitamin B12
e) Thiamine

Q.20. All of the following statements regarding glucose control are true *except* ________.
a) Glucagon stimulates glycogenolysis, gluconeogenesis, and lipolysis.
b) Insulin stimulates glycogen synthesis, gluconeogenesis, and lipolysis.
c) Glucagon stimulates catabolic processes (mobilizes energy) to prevent hypoglycemia.
d) Insulin promotes storage of glucose, fatty acids, and amino acids by their conversion to glycogen, triglycerides, and protein, respectively.
e) Insulin and glucagon exert opposing effects on blood glucose concentrations.

Q.21. An individual exposed to 10 rads (0.1 Gy) of whole body x-irradiation would be expected to ________.
a) have a severe bone marrow depression
b) die
c) be permanently sterilized
d) exhibit no symptoms
e) vomit

Q.22. Benzene is similar to toluene ________.
a) in its metabolism to redox active metabolites
b) regarding covalent binding of its metabolites to proteins
c) in its ability to produce CNS depression
d) in its ability to produce acute myelogenous leukaemia
e) in its ability to be metabolized to benzoquinone

Q.23. Consumption of milk from goats that have grazed on lupine plants containing the alkaloid, anagyrine, may cause ________.
a) birth defects when ingested by women during early pregnancy
b) severe liver damage characterized by centrolobular necrosis
c) dizziness, nausea, headaches, and hallucinations
d) numbness of the extremities
e) aphrodisia and a general increase in sexual awareness

Q.24. Sorbitol and other sugar alcohols have been associated with ________.
a) respiratory distress syndrome
b) osmotic diarrhoea
c) hepatotoxicity
d) immediate hypersensitivity reaction
e) CNS depression

Q.25. Which of the following is *not* regarding Amanita phalloides mushrooms?
a) Toxic components are phalloidin and amatoxins.
b) Produces liver and gastrointestinal toxicity.
c) Cardiovascular toxicity is responsible for mortality.
d) The common name is 'death cap.'
e) No specific antidotal treatment of poisoning is available.

Q.26. Chloroform is *not* ________.
- a) a central nervous system depressant
- b) hepatotoxic
- c) metabolized to phosgene
- d) a peroxisome proliferator
- e) a contaminant of chlorinated water

Q.27. Increasing the casein content of a partially purified diet up to a level of 36% ________.
- a) reduces the spontaneous hepatoma incidence below background in mice
- b) increases the spontaneous hepatoma incidence above background in mice
- c) has no effect on the incidence of mouse hepatoma
- d) reduces the spontaneous incidence of kidney tumours below background in mice
- e) reduces the spontaneous incidence of lung tumours below background in mice

Q.28. All of the following may cause metabolic acidosis *except* ________.
- a) renal failure
- b) salicylates
- c) methanol
- d) diuretics
- e) diarrhoea

Q.29. A patient is admitted to the emergency room with the following symptoms: dry mouth and skin; weak, rapid pulse (130 beats/min); elevated body temperature (103°F); and mydriasis. He is excited and disoriented. In his pocket is a bottle of pills labelled: 'Take one as necessary for stomach pain.' This patient is most likely to be suffering from an overdose of ________.
- a) a narcotic analgesic
- b) a non-narcotic analgesic
- c) an antacid
- d) an antimuscarinic agent
- e) a benzodiazepine tranquilizer

Q.30. Each of the following solvents is paired with a correct target organ of toxicity *except* ________.
- a) methanol:retina
- b) ethylene glycol:kidney
- c) ethylene glycol monomethyl ether:kidney
- d) dichloromethane:central nervous system
- e) carbon tetrachloride:liver

Q.31. In a reproductive toxicity study, what is the 'fertility index'?
- a) The percentage of live fetuses per litter
- b) The percentage of attempted matings that result in pregnancies
- c) The ratio of fetuses at 14 days gestation/total implantations
- d) The ratio of early fetal deaths/total implantations

Q.32. What changes in serum enzyme levels are indicative of acute hepatocellular injury?
- a) Increased alanine aminotransferase and aspartate aminotransferase
- b) Increased lactate dehydrogenase and decreased bilirubin
- c) Increased sorbitol dehydrogenase and lipase
- d) Increased 5′ nucleotidase and decreased urea nitrogen

Answers (Exercise 6)

1. b	4. c	7. e	10. a	13. b	16. b	19. a	22. c	25. c	28. d	31. b
2. c	5. d	8. c	11. c	14. c	17. c	20. b	23. a	26. d	29. d	32. a
3. b	6. a	9. e	12. d	15. e	18. d	21. a	24. b	27. b	30. c	

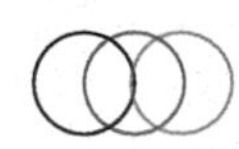

Exercise 7

Q.1. All of the following are true of parathyroid hormone-related proteins *except* ________.
  a) It is present in human tumours.
  b) It plays a role in calcium balance in the fetus.
  c) It has no structural similarity to parathyroid hormones.
  d) It is found in high concentration in milk.

Q.2. A contaminant of drinking water from combustible products that can inhibit the transport of iodide into the thyroid is ________.
  a) sulfuric acid
  b) lead
  c) perchlorite
  d) none of the above

Q.3. Chemicals/drugs that inhibit the organification of thyroglobulin include all of the following *except* ________.
  a) thiourea
  b) sulfonamides
  c) methimazole
  d) furosemide

Q.4. A xenobiotic that inhibits thyroid hormone release is ________.
  a) atrazine
  b) lithium
  c) DDT
  d) arsenic

Q.5. Which of the following statements is *not* correct?
  a) Phenobarbital treatment in rats increases serum TSH by inducing glucuronidation enzymes.
  b) Phenobarbital is a thyroid gland tumour promoter in some studies.
  c) There is a risk of thyroid neoplasms.
  d) Epileptic patients on phenobarbital have an increased risk of liver neoplasms.

Q.6. Mechanism of toxicity to the adrenal cortex include all of the following *except* ________.
  a) activation of the fetal adrenal gene
  b) impaired steroidogenesis
  c) toxin activation by P450 enzymes
  d) lipidosis-producing chemicals

Q.7. Captopril is ________.
  a) an inhibitor of cholesterol ACYL transference
  b) an inhibitor of 21-alpha hydroxylase
  c) an inhibitor of tyrosine hydroxylase
  d) an inhibitor of angiotensin-converting enzyme

Q.8. Adrenal medullary proliferation tumours are called ________.
  a) adrenal adenomas
  b) pheochromocytoma
  c) mesotheliomas
  d) none of the above

Q.9. A xenobiotic associated with causing pituitary tumours in humans is ________.
  a) phenobarbital
  b) bromocriptine
  c) cobalt
  d) none of the above

Q.10. The active transport of an iodide anion can be inhibited by ________.
  a) thiocyanate
  b) bicarbonate

c) lactate
d) acetate

Q.11. A food additive that causes inhibition of 5′-monodeiodonase in the rat is ________.
a) red dye # 3
b) orange B
c) yellow # 5
d) blue # 1

Q.12. Papillary thyroid carcinoma in humans is strongly related to ________.
a) polio virus
b) rifampin
c) radiation exposure
d) TCDD

Q.13. The observed gender differences in thyroid cancer prevalence in humans suggest that ________.
a) androgens play a role in tumorigenesis
b) estrogens play a role in tumorigenesis
c) body weight plays a role in tumorigenesis
d) age plays a role in tumorigenesis

Q.14. Thyroid C cells are known to have receptors for hormones produced in the ________.
a) adrenal medulla
b) central nervous system
c) gastrointestinal tract
d) kidney

Q.15. Calcium concentration in body fluids is regulated by all of the following *except* ________.
a) calcitonin
b) vitamin D
c) parathyroid hormone
d) angiotensin II

Q.16. The most common endocrine organ to be affected by chemicals is the ________.
a) pancreas
b) parathyroid gland
c) adrenal gland
d) pituitary gland

Q.17. Pituitary tumours in rats is caused by ________.
a) hydrocortisone
b) calcitonin
c) angiotensin II
d) none of the above

Q.18. Which of the following is true regarding Leydig cell tumors?
a) They are the most common testicular neoplasm in humans.
b) They are caused by xenobiotics with androgen agonist activity.
c) Ninety percent of human tumours are malignant.
d) The rat is an inappropriate model for assessing the risk of xenobiotic-induced tumours in humans.

Q.19. All the following are true of raloxifene *except* ________.
a) It is associated with increased risk of ovarian cancer in women.
b) It is an estrogen agonist on bone and estrogen antagonist on breast.
c) It increases circulating LH levels in mice.
d) It is associated with ovarian tumour development in mice.

Q.20. What is the target organ of ________?
a) Eye
b) Heart

 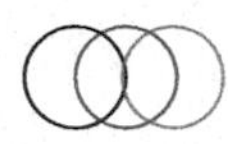

c) Ear
d) Liver

Q.21. A chemical causes cancer in animals but is only positive in the Ames bacterial mutagenicity assay when S9 fraction is added. Is it ________?
a) an ultimate carcinogen
b) a promoter
c) a proximate carcinogen
d) a non-genotoxic carcinogen
e) a co-carcinogen

Q.22. Chemicals that are active during the first week of pregnancy after fertilization of the egg is most likely to cause which effect in the embryo?
a) Death
b) Malformations
c) Functional abnormalities
d) Growth retardation
e) Sterility

Q.23. The drug thalidomide is a teratogen in humans. What is the major malformation it causes?
a) Aphakia
b) Phocomelia
c) Spina bifida
d) Anophthalmia
e) Cleft palate

Q.24. The gestation period in humans is 9 months. What is the equivalent period in the rat?
a) 12 weeks
b) 21 days
c) 45 days
d) 15 days
e) 15 weeks

Q.25. Which of the following causes centrilobular necrosis in the liver due to a reactive metabolite?
a) Ethanol
b) Carbon tetrachloride
c) Bromobenzene
d) Paracetamol
e) b, c, and d

Q.26. Carbon tetrachloride causes fatty liver mainly because of ________.
a) increased uptake of fat from the diet
b) increased secretion of fat from the liver
c) increased mobilization of fat from adipose tissue
d) decreased secretion of fat from the liver
e) decreased metabolism of fat in the liver

Q.27. The hepatocytes in the centrilobular region of the liver ________.
a) are less susceptible to hepatotoxic chemicals than other liver cells
b) have a higher concentration of cytochrome P450 than other hepatocytes
c) have a higher oxygen concentration than those in other regions
d) have the highest glutathione content of any cells in the liver
e) are located in Zone 1

Q.28. A teratogen causes defects usually because of ________.
a) high exposure of the mother prior to fertilization
b) exposure of the female during organogenesis
c) exposure of the male during spermatogenesis
d) exposure of the neonate via the milk
e) none of the above

Q.29. Aplastic anaemia may be caused by ________.
a) benzene exposure
b) fava beans
c) treatment with methyldopa
d) exposure to phenylhydrazine
e) Primaquine

Q.30. Heinz bodies are a manifestation of the toxic effects of a chemical on ________.
a) lipids
b) cytochrome P450
c) haemoglobin
d) the myelin sheath
e) lymphocytes

Q.31. Which of the following is *not* used for the detection of liver damage?
a) Plasma alanine transaminase
b) Urinary conjugated bilirubin
c) Alkaline phosphatase
d) Inulin clearance
e) Plasma albumin/globulin ratio

Q.32. Which of the following is *not* used for the detection of mammalian kidney damage?
a) Urinary y-glutamyltransferase
b) Blood urea nitrogen
c) C-S lyase
d) Glycosuria
e) A 2-microglobulin

Q.33. In most species, a depression in the level of the enzyme cholinesterase in the plasma indicates ________.
a) liver damage
b) myocardial damage
c) organophosphate poisoning
d) kidney damage
e) none of the above

Q.34. In a 90-day chronic toxicity study, it was observed that both red and white blood cell counts were decreased. This is most probably the result of ________.
a) an effect on the spleen
b) an effect on the thymus
c) direct toxicity to blood cells
d) an effect on the bone marrow
e) the induction of leukaemia

Q.35. In drug-induced allergic reactions, which of the following is true?
a) There is no dose response.
b) Only relatively large molecules are normally antigenic.
c) Repeated exposure is necessary.
d) IgE is always involved.

Select the correct statement.
1. 1, 2, and 3
2. 1 and 3
3. 2 and 4
4. only 4
5. all four

Q.36. Adverse effects of drugs in humans may be caused by ________.
1. exaggerated pharmacological effects after overdoses
2. idiosyncratic effects after normal doses

 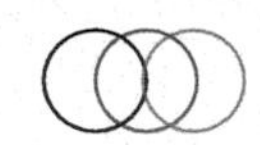

3. toxicity unconnected to pharmacological effect after inappropriate doses
4. dietary constituents

Select the correct statement.

a) 1, 2, and 3
b) 1 and 3
c) 2 and 4
d) only 4
e) all four

Q.37. One response to repeated exposure to chemicals is anaphylactic shock. This ________.

a) results from the production of an antigen
b) leads to a loss of blood pressure
c) causes bronchoconstriction
d) involves the immunoglobulin IgE

Select the correct statement.

a) 1, 2, and 3
b) 1 and 3
c) 2 and 4
d) only 4
e) all four

Answers (Exercise 7)

1. c	5. d	9. d	13. b	17. b	21. c	25. e	29. a	33. a	37. e
2. c	6. a	10. a	14. c	18. d	22. a	26. d	30. c	34. d	
3. d	7. d	11. a	15. d	19. a	23. b	27. b	31. d	35. a	
4. b	8. b	12. c	16. c	20. c	24. b	28. b	32. c	36. e	

4.2 TRUE OR FALSE STATEMENTS

(Write T for True or F for False.)

Exercise 8

Q.1. Toxins affect all organs in the body to the same extent due to their different cell structures.
Q.2. The nephrotoxic effect of most drugs does NOT affect individuals already suffering from kidney failure.
Q.3. Neurotoxins may cause narcosis, behavioural changes, and decreased muscle coordination.
Q.4. Hematopoietic system includes organs other than bone marrow, thymus, lymph nodes, and spleen.
Q.5. The target organ after paraquat toxicity is the kidney.
Q.6. The drug thalidomide taken after delivery can lead to malformations.
Q.7. Toxicant may accumulate in only certain tissues and cause toxicity to that particular tissue, for example, Cd in the kidney.
Q.8. One of the main pathways through which acetaminophen is metabolized is sulfonation (about 52%).
Q.9. Neutral molecules will pass through the glomerulus better than anionic molecules of the same size.
Q.10. The collecting duct has no role in the control of osmolality of urine.

Answers (Exercise 8)

1. F 2. F 3. T 4. F 5. F 6. F 7. T 8. T 9. T 10. F

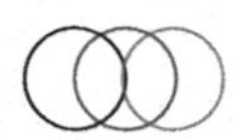

4.3 FILL IN THE BLANKS

Exercise 9

Q.1. The major toxic effect of hydrogen cyanide exposure is ________.

Q.2. Ionizing radiations has the shortest range ________ (i.e., travels the shortest distance in tissue) for the same initial energy.

Q.3. Chloracne is associated with ________.

Q.4. Convulsion in acute poisoning can be treated by ________.

Q.5. Vitamin K is recommended in the treatment of poisoning due to ________.

Q.6. The main target organ of paraquat poisoning is ________.

Q.7. The major organ affected by diquat poisoning is ________.

Q.8. Thalidomide giving birth to deformed children with ________ and ________.

Q.9. After a lot a R&D potential drug candidates fail because of ________.

Q.10. Drug toxicity is also called ________ or ________.

Answers (Exercise 9)

1. Inhibition of mitochondrial respiration
2. Alpha particle
3. Prominent hyperkeratosis of the follicular canal
4. Barbiturates
5. Sweet clover
6. Lung
7. Kidney
8. Shortened limbs and no external ears
9. Toxicity
10. Adverse drug reaction (ADR) or adverse drug event (ADE)

4.4 MATCH THE STATEMENTS

(Column A with Column B)

Exercise 10

Column A		Column B	
Q.1.	indomethacin	a.	spinal bifida
Q.2.	cocaine	b.	staining of teeth
Q.3.	phenytoin	c.	virilization of female fetus
Q.4.	ampicillin	d.	neonatal hypothyroidism
Q.5.	amiodarone	e.	relatively safe
Q.6.	progestins	f.	premature closure of ductus arteriosis
Q.7.	valproic acid	g.	fetal hydantoin syndrome
Q.8.	tetracycline	h.	decreased urine blood flow

Answers (Exercise 10)

1. f 2. h 3. g 4. e 5. d 6. c 7. a 8. b

Exercise 11

Column A		Column B	
Q.1.	oral contraceptives	a.	metal associated with essential hypertension
Q.2.	homocysteine	b.	cardiac hemangiosarcoma in laboratory animals
Q.3.	beta-amyloid	c.	abortion and abruption placentae
Q.4.	carbon disulfide	d.	non-cirrhotic portal hypertension in humans
Q.5.	mercury	e.	increase risk of thrombotic events in users who smoke
Q.6.	particulate matter	f.	preglomerular vasoconstriction and disruption of blood–brain barrier
Q.7.	arsenic	g.	elevated serum levels are associated with increased risk of atherosclerosis and venous thrombosis
Q.8.	cocaine	h.	associated with increased cardiovascular and respiratory morbidity and mortality
Q.9.	lead	i.	may contribute to Alzheimer's disease
Q.10.	1,3-butadiene	j.	endothelial damage and hypothyroidism
Q.11.	hydrazine benzoic acid	k.	smooth muscle tumours in mice

Answers (Exercise 11)

1. e 2. g 3. i 4. j 5. f 6. h 7. d 8. c 9. a 10. b 11. k

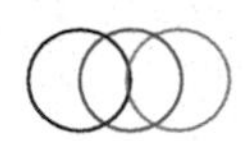

Exercise 12

Column A		Column B	
Q.1.	albumin	a.	elevated in liver disease and haemolysis
Q.2.	aspartate transaminase	b.	elevated in liver and bone disease
Q.3.	prothrombin time	c.	decrease in chronic liver disease
Q.4.	ammonia	d.	elevated in 60%–80% of patients with hepatic encephalopathy
Q.5.	alkaline phosphatase	e.	most sensitive indicator of acute liver disease
Q.6.	ultrasound	f.	distinguishes bone from liver disease
Q.7.	gamma glutamyl transpeptidase	g.	demonstrates extrahepatic bile duct dilation
Q.8.	bilirubin	h.	reflects level of coagulation factors

Answers (Exercise 12)

1. c 2. e 3. h 4. d 5. b 6. g 7. f 8. a

Section 3
NON-ORGAN-DIRECTED TOXICITY

Chapter 5

CARCINOGENICITY AND MUTAGENICITY

5.1 MULTIPLE-CHOICE QUESTIONS

(Choose the most appropriate response.)

Exercise 1

Q.1. Which of the following is characteristic of a non-genotoxic carcinogen?
  a) Has no influence on the promotional stage of carcinogenesis
  b) Would be expected to produce positive responses in in-vitro assays for mutagenic potential
  c) Typically exerts other forms of toxicity and/or disrupts cellular homeostasis
  d) Generally shows little structural diversity
  e) Typically has little effect on cell turnover

Q.2. All of the following are true of epoxide hydrolyases *except* ________.
  a) They add oxygen to a double bond and form a three-member ring.
  b) They are important in hydrolyzing electrophiles.
  c) They play a role in converting benzo(a)pyrene to a carcinogen.
  d) Some forms are inducible.

Q.3. Which of the following is *not* an initiating event in carcinogenesis?
  a) DNA adduct formation
  b) DNA strand breakage
  c) Mutation of proto-oncogenes
  d) Oxidative damage of DNA
  e) Mitogenesis

Q.4. Which of the following is a non-genotoxic liver carcinogen in rats?
  a) Aflatoxin
  b) Vinyl chloride
  c) Pyrrolizidine alkaloids
  d) Clofibrate
  e) Tamoxifen

Q.5. All of the following are true of apoptosis *except* ________.
  a) Cell membrane remains intact.
  b) Early in the process, caspases are activated.
  c) Oxidative stress can initiate it.
  d) It can lead to carcinogenesis.

Q.6. Asbestos exposure is unlikely to cause ________.
  a) lung cancer
  b) GI cancer
  c) emphysema
  d) pulmonary fibrosis
  e) mesothelioma

Q.7. Which of the following is *not* an initiating event in carcinogenesis?
  a) DNA adduct formation
  b) DNA strand breakage
  c) Mutation of proto-oncogenes
  d) Oxidative damage of DNA
  e) Mitogenesis

Q.8. The most potent carcinogen derived from nicotine is ________.
a) naphthene
b) styrene
c) nicotine-derived nitrosamine ketone (NNK)
d) meth tert-butyl ketone

Q.9. Xenobiotic toxicity that occurs after repair and adaptive processes are overwhelmed include all of the following *except* ________.
a) fibrosis
b) apoptosis
c) necrosis
d) carcinogenesis

Q.10. Which of the following is *not* an initiating event in carcinogenesis?
a) DNA adduct formation
b) DNA strand breakage
c) Mutation of proto-oncogenes
d) Oxidative damage of DNA
e) Mitogenesis

Q.11. Which of the following statements is true?
a) Chemical carcinogens in animals are always carcinogens in animals.
b) A chemical that is carcinogenic in humans is usually carcinogenic in at least one animal species.
c) From a regulating perspective, carcinogens are considered to have a threshold dose-response curve.
d) Arsenic is an example of a chemical that is carcinogenic to humans and nearly all species treated.

Q.12. Which of the following is *not* associated with carcinogenesis?
a) Mutation
b) Normal p53 function
c) Ras activation
d) Inhibition of apoptosis
e) DNA repair failure

Q.13. Which of the following species develops cancer after exposure to a PPAR agonist?
a) Monkey
b) Human
c) Mouse
d) Guinea pig

Q.14. All of the following are PPAR agonists *except* ________.
a) 3-metylcholanthrene
b) clofibrate
c) trichloroethylene
d) diethylhexyl phthalate

Q.15. What do PPAR alpha agonists, phenobarbital, and TCDD have in common?
a) They all cause human cancer.
b) They all bind to receptors that bind to response elements that modulate gene transcription.
c) They all bind to nuclear receptors that induce cytochrome 2D6.
d) They are all genotoxic in at least one species.

Q.16. All of the following are agonists at the estrogen receptor in humans *except* ________.
a) DES
b) bisphenol A
c) nonylphenol
d) tamoxifen

 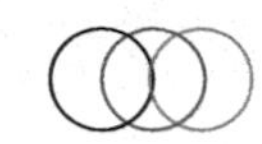

Q.17. Chemicals that increase reactive oxygen species can affect the expression of genes regulating ________.
- a) proliferation
- b) differentiation
- c) apoptosis
- d) all of the above

Q.18. Melamine causes non-genotoxic carcinogenicity by the mechanism of ________.
- a) altered-DNA methylation
- b) induction of oxidative stress
- c) cytotoxicity
- d) stimulation of PPAR alpha-receptors

Q.19. All of the following are true of aromatic amines *except* ________.
- a) they are ultimate carcinogens.
- b) aniline dyes are examples.
- c) they are associated with bladder cancer.
- d) they form reactive metabolites after Phase 1 biotransformation.

Q.20. The carcinogenicity of nongenotoxic chemical that cause cytotoxicity is due to ________.
- a) hormonal factors
- b) increase in spontaneous mutations from secondary hyperplasia
- c) acidosis
- d) active phase reactants

Q.21. All of the following produce renal tumours in male rats *except* ________.
- a) inorganic arsenic
- b) D-limonene
- c) 1,4-dichloribenzene
- d) trimethylpentane

Q.22. PPAR agonists produce all of the following *except* ________.
- a) hepatocellular carcinoma
- b) pancreatic acinar cell tumours
- c) Leydig cell tumours
- d) glioblastoma

Q.23. Which of the following statements are true?
- a) Ethylating carcinogenic agents produce adducts mostly in the phosphate backbone of DNA.
- b) Oxidative DNA adducts occur only on adenine.
- c) The presence of a DNA adduct is sufficient for carcinogenesis.
- d) There is no repair for oxidative DNA damage.

Q.24. All of the following statements are true *except* ________.
- a) The predominate adduct formed from methylating agents on DNA is 7 methyl guanine.
- b) Unpaired electrons on S, O, and N are nucleophilic targets of electrophiles.
- c) Hypomethylated genes are rarely transcribed.
- d) Chemical carcinogen can react with proteins.

Q.25. Which of the following carcinogen-DNA damage pairs is incorrect?
- a) Mustard–DNA cross-links
- b) UV light–pyrimidine dimer
- c) Non-genotoxic carcinogens–7-alkylguanine
- d) Ionizing radiation–double-stranded breaks

Q.26. Which of the following statements is incorrect?
- a) Mutations in an oncogene can result in a clonal cell population with a survival advantage.
- b) DNA damage leads to cell death or neoplasms, never to synthesis to abnormal proteins.

c) Human cancer is usually the result of chronic exposure to a carcinogen.
d) DNA polymerases can correct miscopied DNA bases during replication.

Q.27. All of the following are examples of DNA repair *except* ________.
a) a revere transcriptive repair
b) mismatch repair
c) excision repair
d) end-joining repair of nonhomologous DNA

Q.28. All of the following are reactive carcinogenic electrophiles *except* ________.
a) strained lactones
b) carbonium ions
c) selenium ions
d) epoxides

Q.29. Direct-acting carcinogens include all of the following *except* ________.
a) mustard gases
b) imines
c) sulfate esters
d) aromatic rings

Q.30. Genetically determined abnormal reactivity of an individual to a chemical is known as ________.
a) idiosyncrasy
b) photosensitivity
c) tolerance
d) adaptability
e) none of the above

Answers (Exercise 1)

1. c	4. d	7. e	10. e	13. c	16. d	19. a	22. d	25. c	28. c
2. a	5. d	8. c	11. c	14. a	17. d	20. b	23. a	26. b	29. d
3. e	6. c	9. b	12. b	15. b	18. c	21. a	24. c	27. a	30. a

Exercise 2

Q.1. There is evidence that certain dietary components are carcinogenic. Which of the following is *not* tabbed as a dietary carcinogen?
a) Excessive caloric intake
b) Excessive alcohol consumption
c) Aflatoxin B1 (a food contaminant)
d) Insufficient caloric intake
e) Nitrites (found in some lunchmeats)

Q.2. Which of the following statements regarding mechanisms of chemical carcinogenesis is *false*?
a) Procarcinogens require metabolism in order to exert their carcinogenic effect.
b) Free radicals are highly reactive molecules that have a single, unpaired electron.
c) DNA adducts interfere with the DNA replication machinery.
d) Mutations in the DNA and failure to repair those mutations can be highly carcinogenic.
e) Biological reduction of molecular oxygen is the only way free radicals can be formed.

Q.3. In addition to being necessary for transcription to occur, which of the following transcription factors also plays a crucial role in nucleotide excision repair?
a) FIIA
b) FIIB

 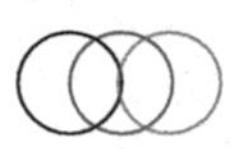

c) FIID
d) FIIF
e) FIIH

Q.4. Which of the following statements regarding DNA repair is true? Base excision repair requires the removal of a longer piece of DNA in comparison with nucleotide excision repair.
a) The repair of double-stranded DNA breaks is more prone to error than is base excision repair.
b) Dimerization of pyrimidines is repaired via base excision repair.
c) Mismatch repair can only recognize normal nucleotides that are paired with a non-complementary nucleotide.
d) Nucleotide excision and base excision are tolerance mechanisms used to respond to DNA damage.

Q.5. Which of the following statements is a characteristic of the initiation stage of carcinogenesis?
a) Initiation is reversible in viable cells.
b) The dose–response exhibits an easily measurable threshold.
c) Cell division is required or the fixation of the process.
d) All initiated cells survive over the lifespan of the organism.
e) Spontaneous initiation of cells is a rare occurrence.

Q.6. Tumour-suppressor genes are mutated in a majority of cancers. Which of the following is *not* characteristic of a tumour-suppressor gene?
a) A mutation in a tumour-suppressor gene is dominant.
b) Germ line inheritance of a mutated tumour-suppressor gene is often involved with cancer development.
c) There is considerable tissue specificity for cancer development.
d) The p53 gene is a tumour-suppressor gene that also acts as a transcription actor.
e) Mutations in tumour-suppressor genes can result in loss of cell-cycle control.

Q.7. Which of the following molecules does *not* play an important role in cell-cycle regulation?
a) p53
b) Cyclin-D
c) MAPK
d) MHC
e) E2F

Q.8. Which of the following environmental factors is proportionally responsible for the *least* amount of cancer deaths?
a) Tobacco
b) Infection
c) Diet
d) Sexual behaviour
e) Alcohol

Q.9. The evidence of the carcinogenicity for dietary intake is sufficient to include one's diet as associated with neoplasms of all of the following *except* ________.
a) colon
b) breast
c) pancreas
d) endometrium
e) gallbladder

Q.10. Which of the following is the correct definition of a complete carcinogen?
a) Chemical capable only of initiating cells
b) Chemical possessing the ability of inducing cancer from normal cells, usually of initiating, promoting, and progression agents

c) Chemical capable of converting an initiated cell or a cell in the stage of promotion to a potentially malignant cell
d) Chemical capable of causing the expansion of initiated cell clones
e) Chemical that will cause cancer 100% of the time that it is administered

Q.11. Which of the following benign–malignant neoplasm pairs is incorrect?
a) Lipoma–liposarcoma
b) Hemangioma–angiosarcoma
c) Squamous cell papilloma–squamous cell sarcoma
d) Bronchial; adenoma–bronchogenic carcinoma

Q.12. All of the following are true of a non-genotoxic carcinogen *except* ________.
a) There is threshold.
b) They are mutagenic.
c) They cause direct DNA damage.
d) They can be tissue specific.

Q.13. All of the following are possible outcomes for initiated cell in the neoplastic process *except* ________.
a) cell death via apoptosis
b) immediate distant metastatic spread
c) can remain in static, non-dividing state
d) can undergo cell division to increase population of initiated cells

Q.14. The progression state of carcinogenesis ________.
a) is reversible
b) involves conversion from preneoplasia to neoplasia
c) does not involve DNA modification
d) always forms a carcinoma

Q.15. All of the following are true of the promotion stage of carcinogenesis *except* ________.
a) multiple cell division is necessary
b) only a single treatment is needed
c) DNA is not directly modified
d) a decrease in apoptosis may be a mechanism

Q.16. The carcinogenicity of inorganic arsenic is unusual in that ________.
a) it causes cancer in humans, but probably not animals
b) it causes different cancers in humans and animal
c) it causes cancer in animals, but not humans
d) it causes cancer in plants, but not in animals

Q.17. The most prevalent oxidative DNA adduct is ________.
a) 5-hydroxyuracil
b) thymine glycol
c) 8-hydroxyguanine
d) 9-oxoguanine

Q.18. Malignant neoplasms of epithelial origin are called ________.
a) sarcoma
b) fibroma
c) carcinomas
d) papilloma

Q.19. Malignant neoplasms of mesenchymal origins are called ________.
a) sarcoma
b) fibroma
c) carcinomas
d) papilloma

Q.20. A carcinogen is an agent that when administered to animals ________.
- a) increases the incidence of malignant neoplasms
- b) increase the incidence of benign neoplasms
- c) increases the incidence of background neoplasms
- d) all of the above

Q.21. An IARC carcinogenic classification of 2A means ________.
- a) the chemical is probably carcinogenic in humans
- b) animal data is positive for cancer development
- c) human epidemiology data is suggestive of cancer causation
- d) all of the above

Q.22. Which of the following occupational carcinogen-cancer-type pairs is incorrect?
- a) Formaldehyde–astrocytoma
- b) Arsenic–skin cancer
- c) Nickel–nasal sinus cancer
- d) Benzidine–bladder carcinoma

Q.23. The development of tumours in rodents after the implantation of solid materials is known as ________.
- a) multiple-hit carcinogenesis
- b) solid-state carcinogenesis
- c) single-hit carcinogenesis
- d) non-mutational carcinogenesis

Q.24. Alpha 2μ globulin nephropathy from hydrocarbons and gastric neuroendocrine cell neoplasia from omeprazole are examples of neoplastic effects in rodents with ________.
- a) no significance to humans
- b) some evidence for human cancer association
- c) strong evidence for human cancer association
- d) IARC 2B classification in humans

Q.25. Cosmetic preparation applied to the skin of the SKH1 albino hairless mouse would likely involve a test for ________.
- a) tumour progression
- b) solid-state carcinogenesis
- c) photochemical carcinogenesis
- d) anti-carcinogenesis

Q.26. The Syrian hamster embryo (SHE) assay is an example of a/an ________.
- a) in-vivo assay
- b) two-year bioassay
- c) organ-specific assay
- d) transformation assay

Q.27. In a classic experimental demonstration of cancer development in mouse kin, croton oil was used as ________.
- a) an initiator
- b) a promoter
- c) a vehicle
- d) a placebo

Q.28. All of the following are of the chronic (two year) carcinogenicity bioassays *except* ________.
- a) FDA mandates the use dogs and monkeys
- b) a vehicle control and two or three doses of test chemicals are used
- c) male and female animals are used
- d) at necropsy, the number, location, and pathology of each tumour are assessed

Q.29. Which of the following lifestyle–cancer associations is incorrect?
- a) Dietary fat–melanoma
- b) Tobacco smoking–bladder cancer

c) Ethanol–oral cancer
d) Moldy food–liver cancer

Q.30. Which of the following drug cancer associations is incorrect?
a) Thorotrast–angiosarcoma of the liver
b) Phenacetin–carcinoma of renal pelvis
c) Diethylstilbestrol–clear cell vaginal carcinoma
d) Estrogens–prostate cancer

Q.31. Which of the following tumour suppressor–gene-neoplasm pairs is incorrect?
a) p16–melanoma
b) Rb1–small-cell lung carcinoma
c) BRCA1–osteosarcoma
d) WT-1–lung cancer

Q.32. All of the following are true of the p53 gene *except* ________.
a) It is essential for checkpoint control during cell division.
b) The active form is a hexamer of six identical units.
c) Enhanced MDM2 in tumour cell decreases functional p53.
d) Mutations are associated with lung, colon, and breast cancer.

Q.33. All of the following are true of carcinogens that inhibit at lower doses and demonstrate protection against carcinogenesis *except* ________.
a) induction of P450 enzyme as a possible mechanism
b) exhibit a J-shaped dose-response curve
c) stimulation of adaptive response that dominate at low doses
d) a mechanism that is only exhibited by tumour promoters

Q.34. Which of the following chemoprotective agent–mechanism pairs is incorrect?
a) Vitamin D–inhibition of cytochrome P450
b) Vitamin C–antioxidant
c) Vanillin–increase DNA repair
d) Folic acid–correct DNA methylation imbalances

Q.35. Big Blue and Muta Mouse are examples of ________.
a) in-vitro gene mutation assays
b) assays that test for tumour promoters and non-initiator
c) transgenic models
d) none of the above

Q.36. All the following statements are true regarding gap-junctional, intracellular communication *except* ________.
a) Small molecules less than 1 Ka can be exchanged through neighbouring cells.
b) It is inhibited by tumour promoters.
c) Carcinogens that interfere with it are not tissue and species specific.
d) It is achieved by connexin hexamers that form a pore between adjacent cells.

Q.37. All of the following statements are true regarding GSTM1 *except* ________.
a) It demonstrates high reactivity toward epoxides.
b) Humans possessing the null isoform have a higher risk for bladder and gastric cancer.
c) It is primarily a detoxication enzyme.
d) The null isoform is protective in breast cancer.

Q.38. All of the following are true regarding proto-oncogenes *except* ________.
a) There are no known oncogenic virus analogues.
b) They are dominant.
c) Somatic mutations can be activated during all stages of carcinogenesis.
d) Germ-line inheritance of these genes is rarely involved in cancer development.

Q.39. All of the following are true regarding oncogenes *except* ________.
a) no known oncogenic virus analogue
b) they are recessive

c) broad tissue specificity for cancer development
d) somatic mutations activated during all stages of carcinogenesis

Q.40. All of the following are true regarding tumour-suppressor genes *except* ________.
a) BCRA1 is an example
b) they are recessive
c) no oncogenic virus analogues
d) germ-line inheritance is never involved in cancer development

Answers (Exercise 2)

1. d	5. c	9. c	13. b	17. c	21. d	25. c	29. a	33. d	37. d
2. e	6. a	10. b	14. b	18. c	22. a	26. d	30. d	34. a	38. a
3. e	7. d	11. c	15. b	19. a	23. b	27. b	31. c	35. c	39. b
4. b	8. e	12. b	16. a	20. d	24. a	28. a	32. b	36. c	40. d

Exercise 3

Q.1. Oncogenes
a) maintain normal cellular growth and development
b) exert their action in a genetically recessive fashion
c) are often formed via translocation to a location with a more active promoter
d) can be mutated to form proto-oncogenes
e) include growth actors and GTPases but not transcription factors

Q.2. Which of the following is NOT one of the more common sources of DNA damage?
a) ionizing radiation
b) UV light
c) electrophilic chemicals
d) DNA polymerase error
e) X-rays

Q.3. Which of the following pairs of DNA repair mechanisms is most likely to introduce mutations into the genetic composition of an organism?
a) Non-homologous end-joining (NHEJ) and base-excision repair
b) Non-homologous end-joining and homologous recombination
c) Homologous recombination and nucleotide-excision repair
d) Nucleotide-excision repair and base-excision repair
e) Homologous recombination and mismatch repair

Q.4. Which of the following DNA mutations would *not* be considered a frameshift mutation?
a) Insertion of 5 nucleotides
b) Insertion of 7 nucleotides
c) Deletion of 18 nucleotides
d) Deletion of 13 nucleotides
e) Deletion of 1 nucleotide

Q.5. Which of the following base-pair mutations is properly characterized as a transversion mutation?
a) T→ C
b) A → G
c) G → A
d) T→ U
e) A → C

Q.6. All of the following statements regarding non-disjunction during meiosis are true *except* ________.
a) Non-disjunction events can happen during meiosis I or meiosis II.
b) All gametes from non-disjunction events have an abnormal chromosome number.

c) Trisomy 21 (Down syndrome) is a common example of non-disjunction.
d) In a non-disjunction event in meiosis I, homologous chromosomes fail to separate.
e) The incorrect formation of spindle fibres is a common cause of non-disjunction during meiosis.

Q.7. Which of the following diseases does *not* have a recessive inheritance pattern?
a) Phenylketonuria
b) Cystic fibrosis
c) Tay-Sachs disease
d) Sickle cell anaemia
e) Huntington's disease

Q.8. What is the purpose of the Ames assay?
a) To determine the threshold of UV light that bacteria can receive before having mutations in their DNA
b) To measure the frequency of aneuploidy in bacterial colonies treated with various chemicals
c) To determine the frequency of a reversion mutation that allows bacterial colonies to grow in the absence of vital nutrients
d) To measure rate of induced recombination in mutagen-treated fungi
e) To measure induction of phenotypic changes in Drosophila

Q.9. In mammalian cytogenic assays, chromosomal aberrations are measured after treatment of the cells at which sensitive phase of the cell cycle?
a) Interphase
b) M phase
c) S phase
d) G1
e) G2

Q.10. Which of the following molecules is used to gauge the amount of a specific gene being transcribed to mRNA?
a) protein
b) mRNA
c) DNA
d) cDNA
e) CGH

Q.11. All of the following statements are true regarding idiosyncratic drug-induced hepatotoxicity *except* ________.
a) It probably involves failure to adapt to a mild drug adverse effect combined with a genetic defect.
b) Traditional animal toxicology studies may not detect it.
c) Carbon tetrachloride is an example.
d) Preclinical studies may need to be done in genetically deficient animals to detect some examples.

Q.12. The FDA protocol that primarily examines fertility and preimplantation and post-implantation viability is ________.
a) Segment I
b) Segment II
c) Segment III
d) Segment IV

Q.13. The FDA protocol that primarily examines postnatal, survival, growth, and external morphology is ________.
a) Segment I
b) Segment II
c) Segment III
d) Segment IV

 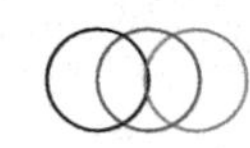

Q.14. All of the following maternal diseases have been assumed with adverse pregnancy outcomes *except* ________.
a) allergic rhinitis
b) febrile illness during the first trimester
c) hypertension
d) diabetes mellitus

Q.15. All of the following are true of the development toxicity of cadmium *except* ________.
a) it appears to involve placental toxicity
b) it appears to involve inhibition of nutrient transport across the placenta
c) zinc can affect the developmental toxicity of cadmium
d) cadmium induces transferrin, which binds zinc in the placenta

Q.16. All of the following are necessary for a normally developing embryo *except* ________.
a) apoptosis
b) cell proliferations
c) cell differentiation
d) necrosis

Q.17. The fetal period is characterized by all of the following *except* ________.
a) beginning organ development
b) tissue differentiation
c) growth
d) physiological maturation

Q.18. Toxic exposure during the fetal period is likely to cause effects on ________.
a) organogenesis
b) implantation
c) growth and maturation
d) all of the above

Q.19. Approximately what percentage of marked drugs belong to the FDA pregnancy category?
a) 1%
b) 10%
c) 20%
d) 30%

Q.20. Diethylstilbestrol (DES) ________.
a) was used to treat morning sickness from the 1940s to the 1970s
b) was found to affect only female off spring in exposed pregnancies
c) greatly affects the development of the fetal brain
d) exposure increases the risk of clear cell adenocarcinoma of the vagina
e) is now used to treat leprosy patients

Q.21. Early (prenatal) exposure to which of the following teratogens is most often characterized by craniofacial dysmorphism?
a) Thalidomide
b) Retinol
c) Ethanol
d) Tobacco smoke
e) Diethylstilbestrol (DES)

Q.22. The nervous system is derived from which of the following germ layers?
a) Ectoderm
b) Mesoderm
c) Epidermal placodes
d) Paraxial mesoderm
e) Endoderm

Q.23. Toxin exposure during which of the following periods is likely to have the *least* toxic effect on the developing fetus?
- a) Gastrulation
- b) Organogenesis
- c) Preimplantation
- d) Third trimester
- e) First trimester

Q.24. Regarding prenatal teratogen exposure, which of the following statements is *false*?
- a) Major effects include growth retardation and malformations.
- b) Exposure to teratogens during critical developmental periods will have more severe effects on the fetus.
- c) There is considered to be a toxin level threshold below which the fetus is capable of repairing itself.
- d) The immune system of the fetus is primitive, so the fetus has little to no ability to fight off chemicals and repair itself.
- e) Embryo lethality becomes more likely as the toxic dose is increased.

Q.25. Which of the following stages of the cell cycle are important in monitoring DNA damage and inhibition?
- a) Progression of the cell cycle
- b) G1–S, anaphase, M–G1
- c) G1–S, S, G2–M
- d) S, prophase, G1
- e) G2–M, prophase
- f) M–G1, anaphase

Q.26. Which of the following molecules is *not* important in determining the ultimate outcome of embryonal DNA damage?
- a) p53
- b) Bax
- c) Bcl-2
- d) c-Myc
- e) NF-κB

Q.27. Which of the following is *not* a physiologic response to pregnancy?
- a) Increased cardiac output
- b) Increased blood volume
- c) Increased peripheral vascular resistance
- d) Decreased plasma proteins
- e) Increased extracellular space

Q.28. All of the following statements are true *except* ________.
- a) Offspring of white mothers have a higher incidence of cleft lip or palate than do black mothers, after adjusting or paternal race.
- b) Cytomegalovirus (CMV) is a common viral cause of birth defects.
- c) Folate supplementation during pregnancy decreases the risk of neural-tube defects.
- d) Cigarette smoke and ethanol are both toxic to the placenta.
- e) In humans, there is a negative correlation between stress and low birth weight.

Q.29. Which of the following is *not* a mechanism involving the endocrine system by which chemicals induce developmental toxicity?
- a) Acting as steroid hormone receptor ligands
- b) Disrupting normal function of steroid hormone-metabolizing enzymes
- c) Disturbing the release of hormones from the hypothalamus
- d) Disturbing the release of hormones from the pituitary gland
- e) Elimination of natural hormones

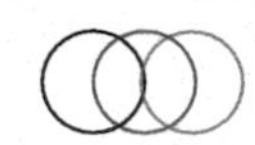

Q.30. Which of the following infectious agents can cause hepatocellular carcinoma?
a) Favivirus
b) Bunyavirus
c) Alphavirus
d) Hepatitis C virus
e) Hepatitis B virus

Q.31. What are the four core tester strains of *Salmonella typhimurium* used in the Ames (bacterial reverse mutation) assay?
a) ta 1535, ta 100, ta 1538, ta 98
b) ta 1540, ta 200, ta 1520, ta 80
c) ta 1548, ta 202, ta 1505, ta 92
d) ta 1536, ta 110, ta 1518, ta 85

Q.32. The in-vivo micronucleus test is often used in a standard battery of genetic toxicity assays. In this test, what do the observed micronuclei represent?
a) Pyknotic nuclei from cells with decreased cytoplasmic to nuclear ratios
b) Clumps of ribosomes and associated RNA fragments
c) Membrane-bounded structures that contain chromosomal fragments or whole chromosomes
d) Binucleated remnants of germ cells

Answers (Exercise 3)

1. c	5. e	9. c	13. c	17. a	21. c	25. b	29. e
2. d	6. b	10. d	14. a	18. c	22. a	26. e	30. e
3. b	7. e	11. c	15. d	19. a	23. c	27. c	31. a
4. c	8. c	12. a	16. d	20. d	24. d	28. e	32. c

5.2 TRUE OR FALSE STATEMENTS

(Write T for True or F for False.)

Exercise 4

Q.1. Carcinogenesis and mutagenesis are synonyms to each other.
Q.2. In carcinogenesis, normal cells are transformed into cancer cells.
Q.3. Tumour-suppressor genes are genes that enhance cell division, survival, or other properties of cancer cells.
Q.4. Random mistakes in normal RNA replication may result in cancer-causing mutations.
Q.5. Viruses that contain an oncogene are categorized as oncogenic because they trigger the growth of tumorous tissues in the host. This process is also referred to as viral transformation.
Q.6. Aneuploidy is the presence of an abnormal number of cancer cells.
Q.7. Chemical agents that damage the genetic information within a cell causing mutations is known as genotoxicity.
Q.8. The skin exposure is considered the most potential for developmental toxicity.
Q.9. Rubella was the first recognized human epidemic of malformations.
Q.10. Sulfonamides were extensively used for the treatment of nausea in pregnant women in late 1950s and early 1960s.

Answers (Exercise 4)

1. F 2. T 3. F 4. F 5. T 6. F 7. T 8. F 9. T 10. F

 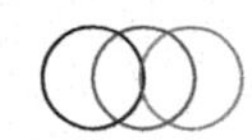

5.3 FILL IN THE BLANKS

Exercise 5

Q.1. Carcinogenesis is also called ________.
Q.2. In carcinogenesis, normal cells are ________ into cancer cells.
Q.3. ________ are genes that inhibit cell division, survival, or other properties of cancer cells.
Q.4. Random mistakes in normal ________ replication may result in cancer-causing mutations.
Q.5. Viruses that contain an oncogene are categorized as oncogenic because they trigger the growth of tumorous tissues in the host. This process is also referred to as ________.
Q.6. Aneuploidy is the presence of an abnormal number of ________.
Q.7. Chemical agents that damage the genetic information within a cell causing mutations is known as ________.
Q.8. The ________ exposure is considered the most potential for developmental toxicity.
Q.9. ________ was the first recognized human epidemic of malformations.
Q.10. ________ was extensively used for the treatment of nausea in pregnant women in late 1950s and early 1960s.

Answers (Exercise 5)

1. Oncogenesis or tumorigenesis
2. Transformed
3. Tumour suppressor genes
4. DNA
5. Viral transformation
6. Chromosomes
7. Genotoxicity
8. First-trimester
9. Rubella
10. Thalidomide

5.4 MATCH THE STATEMENTS

(Column A with Column B)

Exercise 6

Column A		Column B	
Q.1.	neoplasia	a.	do not invade surrounding tissues
Q.2.	benign	b.	functional alteration
Q.3.	genotoxic	c.	birth defects
Q.4.	developmental toxicity	d.	melamine
Q.5.	amelia	e.	autonomous growth of tissue
Q.6.	asbestos	f.	3-metylcholanthrene
Q.7.	PPAR agonist	g.	mutation
Q.8.	melamine	h.	emphysema
Q.9.	non-genotoxic carcinogens	i.	transgenic models
Q.10.	Muta Mouse	j.	7 alkylguanine

Answers (Exercise 6)

1. e 2. a 3. g 4. b 5. c 6. h 7. f 8. d 9. j 10. i

Section 4
TOXIC AGENTS

Chapter 6

TOXIC EFFECTS OF PESTICIDES AND AGROCHEMICALS

6.1 MULTIPLE-CHOICE QUESTIONS

(Choose the most appropriate response.)

Exercise 1

Q.1. Zinc phosphide releases phosphine gas in the following pH:
- a) Acidic
- b) Basic
- c) Neutral
- d) All

Q.2. The following rodenticides require multiple administrations to cause death:
- a) Zinc phosphide
- b) Anticoagulants
- c) Vitamin
- d) Strychnine

Q.3. OP compounds can inhibit the following enzyme(s):
- a) AChE
- b) BuChE
- c) NTE
- d) All

Q.4. The following types of toxicity caused by OP compounds are irreversible even with treatment:
- a) Acute
- b) Subacute
- c) Chronic
- d) OPIDN

Q.5. Which of the following insecticides are more specific to arthropods?
- a) OC compounds
- b) OP compounds
- c) Carbamates
- d) Pyrethroids

Q.6. Paraquat and diquat differ substantially in their _______.
- a) metabolism to a free radical
- b) ability to initiate lipid peroxidation in vivo
- c) uptake by the lung
- d) generation of superoxide anion in vivo
- e) mechanism of cytotoxicity

Q.7. Toxic injury to the cell body, axon, and surrounding Schwann cells of peripheral nerves are referred to, respectively, as _______.
- a) neuropathy, axonopathy, and myelopathy
- b) neuronopathy, axonopathy, and myelinopathy
- c) neuropathy, axonopathy, and gliosis
- d) neuronopathy, dying-back neuropathy, and myelopathy
- e) chromatolysis, axonopathy, and gliosis

Q.8. The following OC insecticides are *not* persistent in the environment:
a) DDT
b) Aldrin
c) Methoxychlor
d) Endosulfan

Q.9. The most commonly used pyrethroid synergist is _______.
a) silica
b) piperonyl butoxide
c) methyl butyl ether
d) *n*-octyl bicyloheptene dicarboximide
e) toluene

Q.10. The chloronicotinyl compound imidacloprid demonstrates a high insecticidal potency and exceptionally low mammalian toxicity due to _______.
a) its high affinity for insect nicotinic acetylcholine receptors and low affinity for mammalian nicotinic acetylcholine receptors
b) the blood–brain barrier in mammals
c) the first-pass effect in the liver in mammals
d) the low pH in the stomach of monogastric mammals
e) the presence of acetylcholinesterase in mammals

Q.11. Fluoroacetate inhibits synthesis of _______.
a) norepinephrine
b) acetylcholine
c) citric acid
d) pyruvic acid

Q.12. Which organic synthetic herbicide is often considered dangerous because it induces accumulation of nitrites in some weed species?
a) Paraquat
b) Glyphosate
c) Lindane
d) 2,4-dichlorophenoxy acetic acid
e) Pentachlorophenol

Q.13. Which category of insecticidal compounds presents a problem of persistent residues in the fatty tissues of animals?
a) Carbamates
b) Organochlorines
c) Organophosphates
d) Pyrethrins
e) Juvenile hormones

Q.14. If acute organophosphate insecticide poisoning is suspected, what is the best initial sample to obtain from a live animal for initial diagnostic testing?
a) Serum
b) Whole blood
c) Urine
d) Stomach contents
e) Fat biopsy

Q.15. Cholinesterase inhibitor pesticides typically cause all of the following *except* _______.
a) salivation
b) miosis
c) dyspnoea
d) blindness
e) bradycardia

 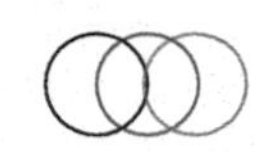

Q.16. When applied to organophosphate insecticide poisoning, the term "aging" refers to _______.
  a) loss of insecticidal activity with time
  b) isomerization of the organophosphate to a more toxic chemical form
  c) hydrolysis of the cholinesterase organophosphate bond induced by oxime drugs
  d) a chemical change that increases the stability of the organophosphate cholinesterase bond
  e) altered toxicity of organophosphates from spontaneous hydrolysis of ester groups

Q.17. Newer anticoagulant rodenticides, also known as second-generation anticoagulants, are important in veterinary medicine because they _______.
  a) have been developed to be toxic in rats but not in other classes of mammals
  b) have effects that are readily treated by synthetic vitamin K injection
  c) are more potent or longer acting than first-generation anticoagulants, requiring prolonged therapy
  d) are more readily detected by chemical analysis than first-generation rodenticides
  e) do not interact with other drugs or chemicals

Q.18. The most-resistant species to DNOC poisoning is _______.
  a) cattle
  b) dog
  c) cat
  d) horse

Q.19. Methoxychlor is less toxic than _______.
  a) DDT
  b) perthane
  c) heptachlor
  d) endrin

Q.20. Inhibition of acetylcholinesterase (AChE) is the mechanism of action of which pesticide?
  a) Organochlorines
  b) Pyrethrins
  c) Organophosphates
  d) Rotenone
  e) Fipronil

Q.21. Which of the following is considered treatment for organophosphate but cannot be used for carbamate poisoning?
  a) Atropine sulfate (muscarinic antagonism)
  b) 2-PAM
  c) Gastric lavage
  d) Activated charcoal
  e) All can be used to treat both organophosphates and carbamate poisoning

Q.22. Nitrogen fertilizers contain which of the following _______.
  a) ammonia
  b) zinc
  c) copper
  d) manganese
  e) molybdenum

Q.23. What is the specific antidote for organochlorine exposure?
  a) Atropine sulfate
  b) 2-PAM
  c) Methylene blue
  d) Physostigmine
  e) None

Q.24. Which of the following statements regarding organochlorines is *false*?

a) Toxicity is neurological.
b) Organochlorines are highly persistent and bioaccumulate.
c) Chronic organochlorine exposure can lead to eggshell thinning and reduced fertility in birds.
d) Mechanism of action is either by inhibition of Na^+ influx and K^+ efflux or inhibiting GABA receptors Organochlorines are highly volatile, and exposure is most likely via inhalation.

Q.25. Which of the following statements is true regarding pyrethrins and pyrethroids?

a) Pyrethrins and pyrethroids are rapidly hydrolysed and metabolized with the majority eliminated after 12–24 hours.
b) Inhalation exposure of pyrethrins and pyrethroids is low.
c) Pyrethrins and pyrethroids are hydrophilic in nature and do not distribute in fat or nervous tissue.
d) Mammalian Na^+ channels are 1,000 times more sensitive than the insect counterpart.
e) Cats do not have adverse reactions to pyrethrins and pyrethroids because they are a more intelligent species and know not to lick this product off a dog.

Q.26. What is the mechanism of action of Fipronil (Frontline)?

a) Fipronil inhibits the oxidation of NADH to NAD^+, which leads to an energy deficiency in cells.
b) Fipronil binds to the membrane lipid phase near the Na^+ channel.
c) Fipronil inhibits Na^+ influx and K efflux.
d) Fipronil noncompetitively binds GABA receptors and blocks Cl^- influx.
e) Fipronil is an AChE inhibitor.

Q.27. Which of the following pesticides is toxic to herding breeds with a mutation in the ABCB1 transporter?

a) Ivermectin
b) Imidacloprid
c) Rotenone
d) Fipronil
e) Pyrethrins

Q.28. Amitraz is an alpha-2 agonist of the CNS that results in sedation. Which drug would be used to treat Amitraz poisoning?

a) Atropine sulfate
b) Yohimbine
c) 2-PAM
d) Atipamezole
e) Both yohimbine and atipamezole

Q.29. Paraquat is a highly toxic herbicide that causes damage through generation of reactive oxygen species. Which organ does paraquat accumulate in?

a) Heart
b) Brain
c) Lung
d) Spleen
e) Small intestine

Q.30. Which of the following statements is *false* regarding Bromethalin?

a) Uncouples oxidative phosphorylation in mitochondria thus depleting ATP
b) Highly lipophilic with brain and fat having high concentrations
c) Causes a build-up of vitamin D3 metabolites, which causes an increase in plasma Ca^+ and P^+
d) Developed for use against warfarin-resistant rodents
e) Secondary poisoning can occur in dogs and cats that eat rodents killed by bromethalin

Q.31. Organochlorine insecticides protect against the acute toxicity of several organophosphorus insecticides by _______.
- a) stimulating enzymatic detoxification of organophosphates
- b) increasing non-catalytic binding sites of organophosphates
- c) stimulating enzymatic detoxification as well as increasing non-catalytic binding sites of organophosphates

Q.32. Ethylene dibromide in high concentrations produces _______.
- a) lung oedema
- b) nephritis
- c) hepatitis

Q.33. Carbamate insecticides interact with _______.
- a) esteratic site of acetylcholine enzyme (AChE)
- b) anionic site of AChE
- c) esteratic as well as anionic site of AChE

Q.34. Red squill is obtained from the plant _______.
- a) *Urginea maritime*
- b) *Strophanthus* sp.
- c) *Croton tiglium*
- d) *Cassia fistula*

Q.35. Cypermethrin-induced toxicity is laboratory animals is also known as _______.
- a) Turner's syndrome
- b) T-syndrome
- c) X-disease
- d) CS Syndrome

Q.36. The diagnostic symptom of chronic pyrethroid toxicity is _______.
- a) muscular twitching
- b) grinding of teeth
- c) hypothermia
- d) irritation

Q.37. The avian toxicant 4-aminopyridine is toxic to dogs and causes clinical effects similar to those of _______.
- a) arsenic
- b) ethylene glycol
- c) lead
- d) organophosphates
- e) strychnine

Q.38. Which of the following reason(s) contribute to the susceptibility of ruminants for urea poisoning?
- a) Urease present in plants
- b) Alkaline pH of rumen
- c) Acidic pH of stomach
- d) Urease activity of rumen

Q.39. Which of the following age groups is more resistant to urea toxicity?
- a) Calf
- b) Heifer
- c) Bull
- d) Cow

Q.40. Which of the following does *not* contribute to the environmental presence of organochlorine insecticides?
- a) High water solubility
- b) Low volatility
- c) Chemical stability

d) Low cost
e) Slow rate of degradation

Q.41. All of the following are characteristic of DDT poisoning *except* _______.
a) paresthesia
b) hypertrophy of hepatocytes
c) increased potassium transport across the membrane
d) slow closing of sodium ion channels
e) dizziness

Q.42. Anticholinesterase agents _______.
a) enhance the activity of AChE
b) increase ACh concentration in the synaptic cleft
c) only target the neuromuscular junction
d) antagonize ACh receptors
e) cause decreased autonomic nervous system stimulation

Q.43. All of the following symptoms would be expected following anticholinesterase insecticide poisoning *except* _______.
a) bronchodilation
b) tachycardia
c) diarrhoea
d) increased blood pressure
e) dyspnoea

Q.44. Which of the following insecticides blocks the electron transport chain at NADH-ubiquinone reductase?
a) Nicotine
b) Carbamate esters
c) Nitromethylenes
d) Pyrethroid esters
e) Rotenoids

Q.45. What is the main mechanism of pyrethroid ester toxicity?
a) Blockage of neurotransmitter release
b) Inhibition of neurotransmitter reuptake
c) Acting as a receptor agonist
d) Causing hyperexcitability of the membrane by interfering with sodium transport
e) Interfering with passive transport across the axonal membrane

Q.46. Which of the following herbicides is *not* correctly paired with its mechanism of action?
a) Glufosinate – inhibition of glutamine synthetase
b) Paraquat – interference with protein synthesis
c) Glyphosate – inhibition of amino acid synthesis
d) Chlorophenoxy compounds – growth stimulants
e) Diquat – production of superoxide anion through redox cycling

Q.47. Captan _______.
a) is an herbicide that inhibits root growth
b) is an insecticide that targets the reproductive organs
c) is a fungicide that could cause duodenal tumours
d) is an herbicide that stimulates growth
e) is a fungicide that is a known teratogen

Q.48. What is a mechanism of action of nicotine?
a) Nicotine antagonizes ACh at the neuromuscular junction.
b) Nicotine decreases the rate of repolarization of the axonal membrane.
c) Nicotine interferes with sodium permeability.
d) Nicotine acts as an ACh agonist in the synapse.
e) Nicotine inhibits the release of neurotransmitter.

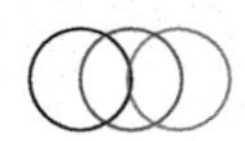

Q.49. Which of the following is the most characteristic of warfarin poisoning?

a) Diarrhoea
b) Cyanosis
c) Decreased glucose metabolism
d) Hematomas
e) Seizures

Q.50. Nitrogen fertilizers contain which of the following _______.

a) ammonia
b) zinc
c) copper
d) manganese
e) molybdenum

Answers (Exercise 1)

1. a	7. b	13. b	19. c	25. a	31. c	37. e	43. a	49. d
2. b & c	8. c & d	14. b	20. c	26. d	32. d	38. a & b	44. e	50. a
3. d	9. b	15. d	21. b	27. a	33. a	39. a	45. d	
4. d	10. a	16. d	22. a	28. e	34. a	40. a	46. b	
5. d	11. c	17. c	23. e	29. c	35. d	41. c	47. c	
6. a	12. d	18. a	24. e	30. c	36. a	42. b	48. d	

6.2 TRUE OR FALSE STATEMENTS

(Write T for True or F for False.)

Exercise 2

Q.1. Sodium arsenate is less toxic than sodium arsenite.
Q.2. Synthetic pyrethroids are more toxic than naturally occurring pyrethrins.
Q.3. Like mammalian species, pyrethroids are less toxic to aquatic organisms.
Q.4. Phosphine released from aluminium phosphide is less acutely toxic than methyl bromide.
Q.5. The specific antidote for diquat poisoning is atropine.
Q.6. Phase I reactions are detoxification reactions.
Q.7. Mitochondrial fraction of the cell is a part of the microsomal fraction.
Q.8. Echothiopate is an organochlorine compound that interacts with both the anionic and esteric sites of AChE to produce a stable complex.
Q.9. Mammals are less susceptible to OP poisoning because of presence of enzyme carboxylesterase.
Q.10. Warfarin is not a coumarone derivative used in rodent control.
Q.11. Oxygen therapy is advocated in cases of diquat poisoning.
Q.12. Fatty and emaciated animals are more sensitive to organochlorine insecticides.
Q.13. ANTU can produce fatal poisoning in dogs when stomach is full.
Q.14. Cholinesterase reactivator that cannot cross the blood–brain barrier is DAM.
Q.15. Extensive use of 2,4-D decreases the nitrate content of barley plants.
Q.16. Zineb breaks down to form ethylene thiourea in the mammalian body.
Q.17. 2-PAM is used as an antidote to treat poisoning due to carbaryl.
Q.18. TCDD is non-toxic.
Q.19. The mechanism of toxicoses of phenoxy herbicides is well understood.
Q.20. Rotenone is obtained from the plant *Chrysanthemum*.
Q.21. Dinitrophenol herbicide imparts a yellowish-green colour to tissues and urine.
Q.22. Red squill is a very good rodenticide.
Q.23. Barbiturate is a drug of choice to control miosis in animals.
Q.24. Pyrethrins are produced from extracts of the plant genus *Nicotiana*.
Q.25. Nicotine is used in the treatment of organophosphorus poisoning.
Q.26. Pentachlorphenol is a safe fungicide due to its poor absorption through intact skin.
Q.27. Fluoroacetate causes citrate accumulation in the cells.
Q.28. Sulfur is an inorganic insecticide.
Q.29. Coumaphos is an OP compound used as an anthelmintic.
Q.30. Deltametharin is classified as a type-I pyrethroid on the basis of chemical nature.
Q.31. Fenthion is a directly acting OP insecticide.
Q.32. Examples of manufactured organic fertilizers include compost, blood meal, bone meal, and seaweed extracts.
Q.33. Oxygen therapy is advocated in cases of diquat poisoning.
Q.34. Use of inorganic fertilizers sometimes lead to deficiency of minerals.
Q.35. Carbon tetrachloride is used for the treatment of fascioliasis.
Q.36. Glyphosate has a narrow margin of safety and is readily absorbed both dermally and orally.
Q.37. Nitrogen fertilizer is often synthesized using the Haber -Bosch process, which produces zinc.
Q.38. Synthetic pyrethroids are more toxic than naturally occurring pyrethrins.
Q.39. Organochlorine insecticides penetrate less through intact skin.
Q.40. Scottish terriers exposed to Phenoxy herbicides will have an increased risk of transitional cell carcinoma of the bladder.
Q.41. Inhibition of acetylcholinesterase (AChE) by organophosphates is reversible.

 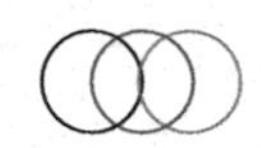

Q.42. The most appropriate treatment for pyrethrin and pyrethroid toxicity is bathing the animal in flea/tick pet shampoo.
Q.43. Mammals are less susceptible to OP poisoning because of the presence of enzyme carboxylesterase.
Q.44. Concentrated sulfuric acid does not char the organic matter.
Q.45. Fluoroacetate causes citrate accumulation in the cells.
Q.46. Hyperthermia is one of the clinical symptoms exhibited in poisoning due to organophosphate compounds.
Q.47. Malathion is less toxic than a paraquat insecticide.
Q.48. Inhibition of acetylcholinesterase (AChE) by carbamates is considered reversible.

Answers (Exercise 2)

1. T	7. F	13. T	19. F	25. F	31. F	37. F	43. T
2. F	8. F	14. F	20. F	26. F	32. T	38. F	44. F
3. T	9. T	15. F	21. T	27. T	33. F	39. F	45. T
4. F	10. F	16. T	22. F	28. F	34. T	40. T	46. F
5. F	11. F	17. F	23. F	29. T	35. T	41. F	47. T
6. F	12. T	18. F	24. F	30. F	36. F	42. F	48. T

6.3 FILL IN THE BLANKS

Exercise 3

Q.1. OC compound insecticides act by inhibiting _______ receptors.

Q.2. Hyperthermia in OC insecticide poisoning is due to changes in metabolism of _______ and _______ neurotransmitters.

Q.3. The metabolite of DDT (dichloro-diphenyl-trichloro-ethane) is _______.

Q.4. DDT acts as an agonist for _______ receptors and DDE acts as an antagonist for _______ receptors.

Q.5. DDE, the metabolite of DDT, causes thinning of egg shells due to inhibition of _______ enzyme.

Q.6. The predominate symptoms in OC insecticide poisoning are _______ and _______.

Q.7. The sedative of choice used in OC insecticide-induced CNS excitation is _______.

Q.8. Death in OC compound poisoning is due to _______ failure.

Q.9. Organic insecticides that are esters of phosphorus are _______ compounds.

Q.10. OP insecticides act as irreversible inhibitors of _______ enzyme.

Q.11. The use of oxime reactivators is contraindicated in _______ insecticide poisoning.

Q.12. Voltage-gated sodium channels are more sensitive for _______ types of pyrethroids.

Q.13. The metabolite of zinc phosphide that is produced in the body is _______. (Hypophosphite is excreted in urine.)

Q.14. Phosphine gas acts as _______ poison.

Q.15. Bait shyness is not observed with _______ rodenticides.

Q.16. Warfarin inhibits clotting factors, which are dependent on _______ for synthesis.

Q.17. Anticoagulant rodenticides decrease vitamin K synthesis through inhibition of _______ enzyme.

Q.18. Capillary damage seen in warfarin rodenticides is due to the presence of _______ chemical moiety.

Q.19. The hematological tests used to confirm poisoning from anticoagulants are _______ and _______.

Q.20. The specific treatment for anticoagulant poisoning is _______.

Answers (Exercise 3)

1. GABA
2. Serotonin and noradrenaline.
3. DDE (dichloro-diphenyl-dichloro-ethylene)
4. Estrogen, androgen
5. Calcium ATPase
6. Behavioral (nervous), hyperthermia
7. Benzodiazepines
8. Respiratory
9. Organophosphate (OP)
10. Acetylcholinesterase
11. Carbamate
12. Two
13. Hypophosphite
14. Protoplasmic
15. Anticoagulant
16. Vitamin K
17. Vitamin K epoxide reductase
18. Benzalactone
19. Clotting time and prothrombin time
20. Phytomenadione (Vitamin K1)

 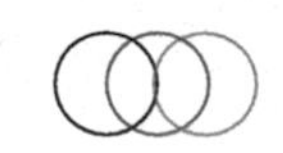

Exercise 4

Q.1. In type II pyrethroid toxicity, characteristic _______ is seen in depolarization phase of action potential.
Q.2. Type II pyrethroids act as inhibitors for _______ neurotransmitter receptors.
Q.3. The symptoms in type I pyrethroid toxicity are referred to as _______.
Q.4. Type II pyrethroids produces a characteristic syndrome known as _______.
Q.5. The group of tranquilizers contraindicated in pyrethroid poisoning are _______.
Q.6. The roots of the *Derris* genus of plants are the source for the natural insecticide _______.
Q.7. Rotenone is highly toxic for _______ species.
Q.8. The agents that are used to control rodents are known as _______.
Q.9. The toxicity of zinc phosphide (Zn_3P_2) is due to the release of _______ gas.
Q.10. The characteristic smell of phosphine gas, which can be used for diagnosis of zinc phosphide toxicity, is _______ odour.
Q.11. The breed of rat, which is very sensitive for ANTU is _______ or _______ rats.
Q.12. The risk of secondary poisoning is very high in _______ rodenticides.
Q.13. Fluoroacetate acts by inhibiting _______ or _______ and lowers energy production in the body.
Q.14. The prominent symptom in fluoroacetate poisoning are _______.
Q.15. The competitive antagonist, which is used in the treatment of fluoroacetate poisoning is _______.
Q.16. The effective rodenticides in warfarin-resistant rats is _______ compounds.
Q.17. The mechanism of rodenticide action of vitamin D compounds (cholecalciferol) is by producing _______.
Q.18. The treatment for vitamin D rodenticides is administration of _______ and _______.
Q.19. The phrase 'drowning in one's own fluids' is associated with _______ rodenticide poisoning.
Q.20. Metaldehyde inhibits _______ neurotransmitter in CNS, which causes excitation.

Answers (Exercise 4)

1. Tail
2. GABA
3. T-syndrome (Tremors)
4. CS syndrome
5. Phenothiazine
6. Rotenone
7. Fish
8. Rodenticides
9. Phosphine (PH_3)
10. Fish-like or Acetylene
11. Brown or norway
12. Second generation
13. TCA or Krebs cycle
14. Neurological
15. Glyceryl monoacetate
16. Vitamin D (cholecalciferol)
17. Hypercalcaemia
18. Corticosteroids and calcitonin
19. ANTU (alpha-naphthyl-thiourea)
20. GABA

Exercise 5

Q.1. Paraquat selectively accumulates in _______ organ of the body.
Q.2. The mechanism of toxicity of paraquat is through generation of _______ free radical, which affects unsaturated membrane lipids.

Q.3. The bipyridyl herbicide, which can cause pulmonary fibrosis is _______.
Q.4. The agents that are used to prevent fungal infestation of plants or seeds are known as _______.
Q.5. Fungicides that can cause act by uncoupling of oxidative phosphorylation are _______.
Q.6. The mechanism of uncoupling of oxidative phosphorylation of PCP is due to _______ activity.
Q.7. Dithio-carbamates, such as ziram and thiram, are derivatives of _______.
Q.8. Agents that are used to control snails and slugs are known as _______.
Q.9. The most commonly used molluscicide is _______.
Q.10. Metaldehyde is extremely toxic through _______ _______ route of exposure.
Q.11. The active metabolite of metaldehyde formed in the body is _______.
Q.12. Organochlorine insecticides such as DDT tend to accumulate in _______ (organ of the body).
Q.13. Organochlorine insecticides such as DDT and BHC tend to be more toxic in oily vehicles due to _______.
Q.14. The group of heterogeneous substances, which are used to control pests, are known as _______ pesticides.
Q.15. Chlorinated hydrocarbons are known as _______ insecticides.
Q.16. The most susceptible species for OC insecticide poisoning is _______.
Q.17. OC insecticides accumulate mainly in _______ tissue of the body.
Q.18. The only OC insecticide that does not accumulate in adipose tissue is _______.
Q.19. The OC compound that is degraded by microbes is _______.
Q.20. The only isomer of benzene hexachloride (BHC), which is biodegradable, is _______ (gamma isomer).
Q.21. OC compounds produce CNS excitation by prolonging _______ phase of action potential.

Answers (Exercise 5)

1. Lungs
2. Superoxide
3. Paraquat
4. Fungicides
5. Pentachlorophenol (PCP)
6. Proton ionophore
7. Carbamates
8. Molluscicides
9. Metaldehyde
10. Inhalation
11. Acetaldehyde;
12. Adipose tissue
13. Increased absorption
14. Pesticides
15. Organochlorine (OC)
16. Cat
17. Adipose
18. Endosulfan
19. Endosulfan
20. Lindane (gamma isomer)
21. Depolarization

Exercise 6

Q.1. OP compounds that inhibit AChE without requiring metabolic activation are called _______.
Q.2. Indirect acting OP compounds require metabolic activation and convert to _______ form before producing toxicity.

Q.3. The toxicity of OP compounds _______ during storage in body tissue.
Q.4. The species of animal that is highly sensitive for OP poisoning is _______.
Q.5. The species of animal that is highly sensitive for delayed neuropathy caused by OP compounds (OPIDN) is _______.
Q.6. The site on AChE with which OP compounds bind and cause irreversible inhibition is _______ site.
Q.7. The phosphorylation of AChE by OP compounds, resulting in the loss of alkyl group and forming an irreversible bond, is called _______.
Q.8. The agents that are used to reactivate OP insecticide-inhibited AChE are called _______.
Q.9. Oxime reactivators are ineffective in reactivating AChE after _______ stage.
Q.10. BuChE (pseudocholinesterase) is present in _______.
Q.11. OPIDN (OP-induced delayed neuropathy) is caused due to inhibition of _______ enzyme.
Q.12. The main symptoms in OP poisoning are _______ symptoms.
Q.13. In OPIDN, the main symptom is _______ type of paralysis.
Q.14. Diagnosis of OP toxicity is carried out by measuring _______ activity in blood.
Q.15. The specific antidote for OP poisoning is _______.
Q.16. Oxime reactivators (2-PAM, DAM) should be used within 24–36 hours' time before aging of _______ occurs.
Q.17. Use of atropine is contraindicated in _______ because it causes cyanosis.
Q.18. Insecticides that cause reversible inhibition of AChE are _______.
Q.19. Carbamates bind with AChE at _______ and _______ sites.
Q.20. Oxime reactivators are ineffective in reactivating AChE in _______ insecticide poisoning.

Answers (Exercise 6)

1. Direct-acting OP compounds
2. Oxon
3. Increases
4. Cat
5. Chicken
6. Esteratic
7. Aging
8. Oxime reactivators
9. Aging
10. Plasma
11. Neurotoxicesterase (NTE)
12. Cholinergic
13. Ascending flaccid
14. AChE
15. Atropine
16. AChE
17. Cats
18. Carbamates
19. Anionic and esteratic
20. Carbamate

Exercise 7

Q.1. Bait shyness and tolerance develops very rapidly for rodenticide _______.
Q.2. The main symptom in ANTU poisoning is _______.
Q.3. The rodenticide red squill is obtained from the plant _______.
Q.4. The toxic glycosidic principle present in red squill is _______.
Q.5. Scillirosides are present in highest concentration in _______ part of the plant.

Q.6. The action of GIT microflora on scilliroside glycosides releases the aglycon portion _______.
Q.7. The action of scillirosides glycosides is similar to _______.
Q.8. The most widely used herbicide is _______.
Q.9. _______ are used to control weeds.
Q.10. Herbicides enhance the palatability of certain toxic plants by increasing the content of _______.
Q.11. _______ is an OP compound used as an anthelmintic.
Q.12. Deltamethrin is classified as _______ on the basis of the chemical nature.
Q.13. Major route of excretion of DDT in cattle is through _______ which cause health hazards in human beings.
Q.14. _______ is a dinitroaniline compound, which is used as herbicide.
Q.15. DDT was discovered and studied by _______, while organophosphorus insecticides by _______.
Q.16. _______ is used to treat rodenticide poisoning because of its anticoagulant action.
Q.17. The species of animal that is more sensitive for 2, 4-D is _______.
Q.18. The manufacturing contaminants that make 2, 4-D extremely toxic are _______.
Q.19. The herbicides that are potent uncouplers of oxidative phosphorylation are _______.
Q.20. Urine is chrome-yellow in colour and turns black upon exposure to air in _______.
Q.21. The most toxic among the herbicides are _______.

Answers (Exercise 7)

1. ANTU
2. Pulmonary oedema
3. *Urginea maritima*
4. Scilliroside
5. Bulb
6. Scillirosidin
7. Digitalis
8. Herbicides
9. 2, 4-D (2, 4-dichloro-phenoxy-acetic acid)
10. Nitrates
11. Coumaphos
12. Type-II pyrethroid
13. Milk
14. Pendimethalin
15. Paul Muller, Gerhard Schrader
16. Vitamin K
17. Dog
18. Dioxins
19. Dinitrophenols
20. Dinitrophenol
21. Bipyridyl group (e.g., paraquat, diquat)

Exercise 8

Q.1. The least-toxic carbamate used in veterinary practice is _______.
Q.2. The specific antidote for carbamate poisoning is _______.
Q.3. Pyrethroids are natural insecticides obtained from _______ flowers.
Q.4. The most widely used household insecticides are _______.
Q.5. Type I pyrethroids resemble _______ in their structure and activity.
Q.6. Type II pyrethroids contain the group _______ in their structure and are more active and toxic.
Q.7. The first commercially available pyrethroid used to date for repelling mosquitoes is _______.

 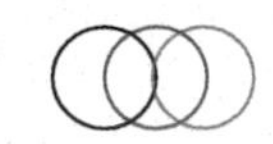

Q.8. The species that is highly susceptibility to pyrethroids toxicity is _______.
Q.9. The relationship between toxicity of pyrethroids and temperature is _______.
Q.10. Pyrethroids cause CNS stimulation through inhibiting the closure of _______ channels.
Q.11. Excessive ammonia inhibits _______ cycle in the body, decreasing energy production.
Q.12. _______ can be a serious problem because items such as rubber bands, socks, rocks, and string can severely damage or block your dog's intestines.
Q.13. The treatment in urea poisoning includes infusion of 5% _______ and _______ into the rumen.
Q.14. Intravenous infusions containing _______ are contraindicated in urea poisoning due to the induction of hyperglycaemia.
Q.15. The toxic principle produced from the interaction of ammonia with reducing sugars in feeds is _______.
Q.16. 4-Methyl imidazole (4-MI) present in ammoniated feeds produces CNS symptoms in cattle known as _______.
Q.17. Urea is added to the diets of ruminants as a source of _______.
Q.18. The lethal dose of urea in cattle is _______.
Q.19. Urea can be used in the diets of ruminants and horses due to the presence of _______ microbial enzyme in the rumen and caecum, respectively.
Q.20. The feed component that is required for efficient utilization of urea is _______.
Q.21. The recommended ratio of urea to molasses for straws and other high-fibre diets is _______.
Q.22. In horses, the site of conversion of urea to ammonia is _______.
Q.23. The nervous symptoms found in salt poisoning are due to the development of _______ in brain.
Q.24. Dragging of hind limbs and knuckling of fetlock joints is a characteristic feature of _______ poisoning in cattle.

Answers (Exercise 8)

1. Carbaryl
2. Atropine
3. *Chrysanthemum*
4. Synthetic pyrethroids
5. Natural pyrethrins
6. α-cyano (e.g. Deltamethrin)
7. Allethrin
8. Fish
9. Inverse
10. Voltage-gated sodium
11. TCA/Krebs
12. Pica
13. Acetic acid (vinegar) and cold water
14. Glucose/dextrose
15. 4-methyl imidazole (4-MI)
16. Bovine Bonker's syndrome
17. Nitrogen
18. 1–1.5 g/day/animal
19. Urease
20. Carbohydrates or total digestible nutrients (TDN)
21. 1:5
22. Caecum
23. Cerebral oedema
24. Salt

6.4 MATCH THE STATEMENTS

(Column A with Column B)

Exercise 9

Column A		Column B
Q.1.	OP compound	a. Hemlock
Q.2.	organochlorine compound	b. Pyrethrin
Q.3.	pyrethroid synergist	c. Endosulfan
Q.4.	*Conium maculatum*	d. Hen Test
Q.5.	OP-induced delayed neurotoxicity	e. Piperonyl butoxide
Q.6.	Amitraz	f. serum
Q.7.	bipyridyl herbicide	g. miosis
Q.8.	rodenticide	h. paraquat
Q.9.	diagnosis	i. Ziram
Q.10.	dithio-carbamates	j. Warfarin
Q.11.	zinc phosphide	k. neonics
Q.12.	paraquat	l. ureas
Q.13.	ammonia	m. alpha 2 agonist
Q.14.	natural insecticide	n. nicotinic
Q.15.	Amitraz	o. citric acid
Q.16.	herbicides	p. lungs
Q.17.	Tabun	q. rotenone
Q.18.	neonicotinoids	r. rat poison
Q.19.	fluoroacetate	s. nerve agent
Q.20.	cholinergic receptors	t. fertiliser

Answers (Exercise 9)

1. g	2. c	3. e	4. a	5. d	6. b	7. h	8. j	9. f	10. i
11. r	12. p	13. t	14. q	15. m	16. l	17. s	18. k	19. o	20. n

CHAPTER 7

TOXIC EFFECTS OF METALS AND MICRONUTRIENTS

7.1 MULTIPLE-CHOICE QUESTIONS

(Choose the most appropriate response, it may be one, two or more.)

Exercise 1

Q.1. Exposure to fumes of which of the following metals is most likely to cause acute chemical pneumonitis and pulmonary edema?
- a) Lead
- b) Zinc
- c) Cadmium
- d) Copper
- e) Magnesium

Q.2. Deficiency of which element in the sow predisposes baby pigs to toxicosis by injectable iron preparations?
- a) Copper
- b) Chromium
- c) Magnesium
- d) Selenium
- e) Zinc

Q.3. Which of the following is *not* commonly associated with mercury-vapour poisoning?
- a) Acute, corrosive bronchitis
- b) Interstitial pneumonitis
- c) Tremor
- d) Increased excitability
- e) Vomiting and bloody diarrhoea

Q.4. Which form of mercury was the predominant cause of Minamata Bay disease?
- a) Metallic mercury
- b) Mercuric salts
- c) Mercurous salts
- d) Organic mercury compounds
- e) Mercury was not the causative agent

Q.5. Which is the only arsenical that can cause blindness?
- a) Arsenic trioxide
- b) Arsenic pentoxide
- c) Arsine
- d) Arsanilic acid

Q.6. Copper has an inverse interrelationship with the following element(s) _______.
- a) iron
- b) molybdenum
- c) sulfur
- d) both b and c

Q.7. In lead poisoning, basophilic stipplings (BS) are commonly seen in this species _______.
 a) cattle
 b) sheep
 c) dog
 d) horse

Q.8. Which of the following chelating agent(s) is/are used for treating mercury poisoning?
 a) Dimercaprol (BAL)
 b) D-Penicillamine
 c) DMSA (Succimer)
 d) Na-thiosulfate.

Q.9. Which of the following nutrient(s) can counteract the toxicity of organic mercurial?
 a) Vitamin A
 b) Vitamin D
 c) Vitamin E
 d) Selenium

Q.10. Which is the only arsenical that can cause blindness?
 a) Arsenic trioxide
 b) Arsenic pentoxide
 c) Arsine
 d) Arsanilic acid

Q.11. Which of the following forms of Hg is more toxic?
 a) Elemental
 b) Monovalent
 c) Divalent
 d) Organic

Q.12. Mercury can cross the following barriers in the body _______.
 a) blood–brain barrier (BBB)
 b) placental barrier (PB)
 c) both
 d) no barrier

Q.13. The following properties can be attributed to methylmercury (organic Hg) _______.
 a) mutagenic
 b) carcinogenic
 c) embryotoxic
 d) both b and c

Q.14. What form of mercury was the cause of Minamata Bay disease?
 a) Mercuric salts
 b) Mercurous salts
 c) Organic mercury
 d) Elemental mercury

Q.15. The concentration of arsenic in the liver/kidney associated with toxicity is _______.
 a) 3–5 ppm
 b) 10–20 ppm
 c) 20–50 ppm
 d) none of above

Q.16. Inorganic arsenic toxicosis is manifested clinically as _______.
 a) icterus, anaemia, and haemoglobinuria
 b) anurosis, incoordination, and constipation
 c) cardiomyopathy, hydrothorax, and ascites
 d) photosensitization, dermatitis, and hair loss
 e) vomiting, gastroenteritis, diarrhoea, and dehydration

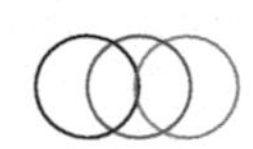

Q.17. Which combination of mineral additives is most useful in preventing chronic copper toxicosis in sheep?
- a) Selenium and molybdenum
- b) Selenium and sulfate
- c) Zinc and molybdenum
- d) Sulfate and molybdenum
- e) Arsenic and sulfate

Q.18. Which of the following form of phosphorus is toxic?
- a) White
- b) Red
- c) Yellow
- d) Black

Q.19. Fluoride inhibits pyruvic-acid synthesis by inhibiting an enzyme _______.
- a) enolase
- b) transaminase
- c) phosphatase
- d) phosphodiesterase

Q.20. Which of the following conditions represents a manifestation of moderate chronic fluoride toxicity?
- a) Osteomalacia
- b) Osteosclerosis
- c) Osteopetrosis
- d) Osteopenia
- e) Osteolysis

Q.21. Which of the following is *not* a major excretory pathway of metals?
- a) Sweat
- b) Urine
- c) Respiration
- d) Feces
- e) Hair

Q.22. Metallothioneins _______.
- a) are responsible or metal transport in the bloodstream
- b) are involved in the biotransformation of metals
- c) invoke hypersensitivity reactions
- d) provide high-affinity binding of copper and mercury
- e) are involved in extracellular transport of metals

Q.23. Which of the following metal-binding proteins is *not* correctly paired with the metal it binds?
- a) Transferring–iron
- b) Ceruloplasmin–copper
- c) Metallothioneins–zinc
- d) Ferritin–lead
- e) Albumin–nonspecific metal binding

Q.24. Which of the following groups is *least* likely to chelate metals?
- a) –COOH
- b) –Cl
- c) –NH
- d) –OH
- e) –SH

Q.25. What is the mechanism of toxicity of arsenic (As)?
- a) Inhibition of mitochondrial respiration
- b) Impairment of calcium uptake by membrane transporters

c) Accumulation in renal corpuscle
d) Abolition of sodium–potassium gradient
e) Destruction of surfactant in the lungs

Q.26. Lead's toxicity is largely due to its ability to mimic and interfere with normal functioning of which of the following ions?
a) Na^+
b) K^+
c) Cl^-
d) Fe^{2+}
e) Ca^{2+}

Q.27. Which of the following statements regarding mercury (Hg) toxicity is *false*?
a) A major source of environmental mercury is rainwater.
b) Mercury vapour is much more dangerous than liquid mercury.
c) Mercury vapour inhalation is characterized by fatigue and bradycardia.
d) Microorganisms in bodies of water can convert mercury vapour to methylmercury.
e) Methylmercury is the most important source of human mercury toxicity.

Q.28. Which of the following is a common symptom of nickel exposure?
a) Renal failure
b) Diarrhoea
c) Hepatic cirrhosis
d) Contact dermatitis
e) Tachycardia

Q.29. Which of the following statements regarding Wilson's disease is *false*?
a) Serum ceruloplasmin is high.
b) Urinary excretion of copper is high.
c) There is impaired biliary excretion of copper.
d) The disease can be treated with liver transplantation.
e) This is an autosomal recessive disorder.

Q.30. Which of the following statements regarding metals and medical therapy is *false*?
a) There are elevated levels of aluminum in the brains of Alzheimer's patients.
b) Lithium is used to treat depression.
c) Chronic nephrotoxicity is a common result of excess aluminum exposure.
d) Platinum is used as cancer treatment.
e) Platinum salts can cause an allergic dermatitis.

Q.31. Alopecia and diarrhoea are the prominent symptoms in poisoning with _______.
a) Copper
b) molybdenum
c) *Leucaena leucocephala*
d) two of above
e) none of above

Q.32. Band formation in the hooves is seen in poisoning due to _______.
a) Fluorides
b) Ergotism
c) Lead
d) Selenium

Q.33. BAL is the specific antidote against poisoning by _______.
a) Selenium
b) Copper
c) Lead
d) Lewisite

 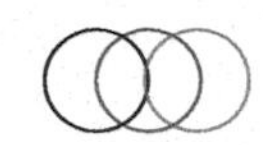

Q.34. The following form of selenium is/are soluble _______.
  a) elemental (0)
  b) selenide (+2)
  c) selenite (+2)
  d) selenate (+6)
  e) organo selenium

Q.35. A group of swine shows paralysis, hoof (corollary band) and hair lesions, and lesions of focal symmetric poliomyelomalacia. The most likely cause of these signs is toxicosis involving _______.
  a) arsenic
  b) copper
  c) lead
  d) selenium
  e) zinc

Q.36. In cattle, chronic fluoride toxicosis causes _______.
  a) diarrhoea, pale hair coat, lameness, and hoof overgrowth
  b) icterus, haemoglobinuria, and photosensitization
  c) emaciation, hair loss, and lameness
  d) rumen stasis, nephrosis, and constipation
  e) lameness, exostosis, and excessive dental wear

Q.37. Fluoride interferes with the following element(s) in the body _______.
  a) calcium
  b) magnesium
  c) manganese
  d) phosphorus

Q.38. Which form of fluorosis is seen if the animal is exposed to fluorine during early stages of life?
  a) Skeletal
  b) Dental
  c) Both
  d) Not affected

Q.39. Which of the following form of phosphorus is toxic?
  a) White
  b) Red
  c) Yellow
  d) Black

Answers (Exercise 1)

1. c	7. c	13. a, b, c & d	19. a	25. a	31. c	37. a, b & c
2. d	8. a, b, c & d	14. c	20. b	26. e	32. d	38. c
3. e	9. c & d	15. a	21. c	27. c	33. d	39. a & c
4. d	10. d	16. e	22. d	28. d	34. c, d & e	
5. d	11. d	17. d	23. d	29. a	35. d	
6. d	12. c	18. a & c	24. b	30. c	36. e	

7.2 TRUE OR FALSE STATEMENTS

(Write T for True or F for False.)

Exercise 2

Q.1. The oxides of metals tend to be toxic, but the oxides of non-metals tend to be non-toxic.
Q.2. A major site of toxic action for metals is interaction with peptides.
Q.3. Metallothionein is a low-molecular-weight metal-binding protein.
Q.4. Mercury is known to produce testicular injury after acute exposure.
Q.5. Cadmium can be found in the circulatory system bound to metallothionein (MT) and form a complex CdMT.
Q.6. Metallic arsenic is not poisonous because it is insoluble in water and cannot be absorbed from the GI tract.
Q.7. The term 'itai-itai disease' was coined by locals for the severe pains felt in the lungs and kidney of victims.
Q.8. Deferoxamine (desferrioxamine) is the specific antidote for iron toxicity.
Q.9. Arsenic is rapidly detoxified and is completely eliminated in few days.
Q.10. The half-life of cadmium is about 24 hours.

Answers (Exercise 2)

1. F 2. F 3. T 4. F 5. T 6. T 7. F 8. T 9. T 10. F

 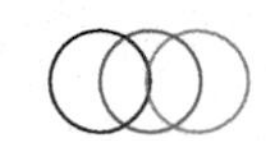

7.3 FILL IN THE BLANKS

Exercise 3

Q.1. In selenium toxicity, depleted glutathione levels can be restored by administering _______ during treatment.
Q.2. Bran disease in horses is caused due to excess feeding of wheat bran, which contains a high amount of _______.
Q.3. In selenium poisoning, the level of selenium detected in blood is _______ and in hooves is _______.
Q.4. A garlic-like odour of breath or stomach contents luminous in dark is indicative of _______ poisoning.
Q.5. Phosphorus is eliminated from the body through _______ and _______.
Q.6. Dermal exposure to white/yellow phosphorus leads to _______.
Q.7. The use of _______ metal chelator is contraindicated in selenium toxicity.
Q.8. The immediate symptom upon oral ingestion of phosphorus is _______.
Q.9. The organs that are damaged in phosphorus poisoning are _______ and _______.
Q.10. In chronic phosphorus poisoning, the necrosis of the jaw that is observed is called _______.
Q.11. The best material for the diagnosis of phosphorus poisoning is _______ or _______.
Q.12. The prognosis in the case of phosphorus poisoning is _______ to _______.
Q.13. Cyanide inhibits cellular respiration by binding with _______.
Q.14. The conversion of methemoglobin to haemoglobin by methylene blue depends on the availability of _______.
Q.15. The detection of opaque lines in metaphysis of bones in an X-ray suggests _______.
Q.16. In lead poisoning, the estimation of _______ enzyme in blood is of diagnostic value.
Q.17. The lead content in the liver and kidney indicative of lead poisoning is more than _______.
Q.18. The specific antidote for lead toxicosis is _______.
Q.19. Meat from food animals recovered from lead poisoning is _______ for human consumption.
Q.20. The species of animals that are most susceptible to lead poisoning are _______, _______ and _______.
Q.21. The species of animal that is considered an indicator for lead in the environment is _______
Q.22. The species of animal that is very resistant to lead poisoning is _______.
Q.23. The most common route of lead exposure is _______.
Q.24. Roaring in horses due to lead poisoning is caused due to _______ nerve paralysis.

Answers (Exercise 3)

1. Acetyl cysteine
2. Phosphorus
3. 1–4 PPM, 5–20 PPM
4. Phosphorus
5. Urine and breath
6. Skin burns
7. Dimercaprol (BAL)
8. Emesis
9. Liver, kidney
10. Phossy jaw
11. Vomitus or stomach contents
12. Guarded to grave
13. Cytochrome oxidase
14. $NADPH_2$
15. Lead

16. ALA-D synthase
17. >4 ppm
18. Calcium disodium EDTA
19. Fit
20. Dog, cattle, and horses
21. Dog
22. Swine
23. Oral route
24. Recurrent laryngeal

Exercise 4

Q.1. The predominant symptom in inorganic mercury poisoning is _______.
Q.2. The predominant symptom in elemental mercury poisoning is _______.
Q.3. Elemental mercury (Hg) is toxic only through _______ route of exposure.
Q.4. The most frequently encountered heavy-metal poisoning in veterinary cases is _______.
Q.5. The sample of choice for detection of inorganic mercury intoxication is _______.
Q.6. The sample of choice for detection of organic mercury intoxication is _______.
Q.7. Neurological and renal damage caused in mercury poisoning are _______.
Q.8. The lead compound that is added to petrol and gasoline as anti-knocking is _______.
Q.9. The lead compound that was used for sweetening of wine is _______.
Q.10. The carcass of animals affected with mercury poisoning is _______ for human consumption.
Q.11. Elemental mercury is toxic when exposed through _______ route.
Q.12. The common source of mercury poisoning in animals is through _______.
Q.13. Arsenic tends to accumulate in _______ and _______ (name the organ of the body).
Q.14. Minamata disease in Japan is caused due to the consumption of fish contaminated with _______.
Q.15. The species of animal that is more sensitive to Hg poisoning is _______.
Q.16. In the body, lead tends to accumulate in tissues such as _______ and _______.
Q.17. The process of accumulation of Hg in marine animals to a very high concentration over a period of time is known as _______.
Q.18. In the body, heavy metals such as mercury and cadmium tend to accumulate in _______ (name the organ of body).
Q.19. The mechanism of toxicity of mercury involves binding with _______, _______ groups of proteins and enzymes.
Q.20. The predominant symptoms of organic mercury poisoning are _______.
Q.21. Acute zinc deficiency causes _______ in children.
Q.22. The common source of mercury poisoning in animals is through _______.

Answers (Exercise 4)

1. Gastroenteritis
2. Pulmonary symptoms
3. Inhalation route
4. Lead poisoning or plumbism
5. Urine
6. Kidney
7. Irreversible
8. Tetraethyl lead
9. Lead acetate
10. Unfit
11. Inhalation
12. Food
13. Hair and skin

 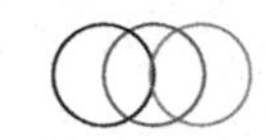

14. Methylmercury
15. Cattle (cow and calves)
16. Bone and teeth
17. Bioaccumulation or biomagnification
18. Kidney
19. –SH, thiol
20. Neurological
21. Erosive dermatitis
22. Food.

Exercise 5

Q.1. Death in fluoride toxicity is due to _______ and _______.
Q.2. The chronic disease that occurs due to continuous ingestion of small doses of fluoride is _______.
Q.3. In the case of fluorosis, treatment with _______ salts is usually followed.
Q.4. Supplementation with _______ mineral supplements can cause fluorosis.
Q.5. The fatal toxicosis that can occur from gases and dust from volcanic eruptions is _______.
Q.6. Acute fluoride poisoning is common in _______ whereas chronic poisoning is common in _______ species of animals.
Q.7. The maximum tolerable level of fluoride in forage for herbivorous animals is _______.
Q.8. The level of fluoride in drinking water, which can cause fluorosis in animals, is _______.
Q.9. In the body, fluoride accumulates in _______ and _______.
Q.10. Fluoride is gradually excreted from the body through _______.
Q.11. If haemolytic crisis develops in copper poisoning, the prognosis is _______.
Q.12. Fluoride causes the formation of _______ in the acidic medium of the stomach, which is responsible for the corrosive action.
Q.13. Hyperkalemia occurring in fluoride poisoning is due to the inhibition of _______ enzyme.
Q.14. Brown or black discolouration of teeth seen in fluorosis is due to oxidation of _______, which is defective in the tooth.
Q.15. In fluorosis, defects in bones are due to replacement of _______ groups with fluoride in hydroxyapatite structure.
Q.16. Chronic fluorosis is manifested in _______ and _______ forms.
Q.17. The prognosis in case of phosphorus poisoning is _______.
Q.18. The samples of choice to be collected in suspected cases of fluoride poisoning are _______ and _______.
Q.19. In fluorosis, the density of bones is _______.
Q.20. The soft tissue that accumulates highest amount of fluorine in the body is _______.

Answers (Exercise 5)

1. Hyperkalemia, hypocalcaemia
2. Fluorosis
3. Calcium
4. Rock phosphate
5. Fluorine intoxication
6. Dogs; herbivores
7. 40–50 ppm
8. >2 ppm.
9. Bone, teeth
10. Urine
11. Grave
12. Hydrofluoric acid

13. Na^+-K^+ ATPASE;
14. Enamel
15. Hydroxyl
16. Skeletal and dental
17. Guarded to grave
18. Bone and urine
19. Increased
20. Pineal gland

Exercise 6

Q.1. The type of climates in which selenium toxicity is more common is _______ climates.
Q.2. The pH of the soil that favours selenium accumulation is _______.
Q.3. The metabolite of selenium excreted through urine is _______.
Q.4. Cytotoxicity of selenium is due to the generation of _______ at the cellular level.
Q.5. The non-enzymatic antioxidant depleted by selenium is _______.
Q.6. In selenium poisoning, the characteristic odour observed is _______.
Q.7. The vitamin that can aggravate selenium poisoning is _______.
Q.8. Subacute toxicity of selenium is also known as _______.
Q.9. Chronic selenium toxicity is also called as _______.
Q.10. In cattle, cracked and overgrown hooves are the main symptoms of _______ poisoning.
Q.11. In horses, loss of hair from the mane is the main symptom of _______ poisoning.
Q.12. The most toxic from of selenium is _______.
Q.13. Organo-selenium is formed due to the replacement of _______ by selenium in amino acids.
Q.14. The level of selenium in plants that is toxic to animals is _______.
Q.15. The type of plants that physiologically require selenium and accumulate high levels of selenium are called _______.
Q.16. The level of selenium found in obligate accumulators is _______.
Q.17. The types of plants that do not require selenium but accumulate high levels of selenium if present in soil are called _______.
Q.18. The level of selenium found in facultative accumulators is _______.
Q.19. The types of plants that do not require selenium but accumulate low levels of selenium if present in soil are called _______.
Q.20. The level of selenium found in non-accumulators is _______.

Answers (Exercise 6)

1. Arid and semi-arid
2. Alkaline (>7.0)
3. Trimethyl-selenonium
4. Free radicals
5. Glutathione (GSH)
6. Garlic-like
7. Vitamin E
8. Blind staggers
9. Alkali disease
10. Selenium
11. Selenium
12. Organo selenium
13. Sulfur
14. 5 ppm and above
15. Obligate (primary) accumulators or indicator plants

 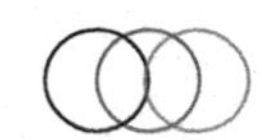

16. 100–1500 ppm
17. Facultative accumulators
18. 25–100 ppm
19. Non-accumulators
20. 1–25 ppm.

Exercise 7

Q.1. The chelating agent used to treat copper poisoning is _______.
Q.2. The species of animal that is more susceptible of copper poisoning is _______.
Q.3. The species that is highly resistant to copper poisoning is _______.
Q.4. The ratio of copper to _______ in feeds should ideally be 6:1.
Q.5. The breed of dog that is highly susceptible to copper toxicosis due to genetic predisposition is _______.
Q.6. The deficiency of _______ micromineral predisposes to copper toxicity.
Q.7. The specific transport proteins for copper in the body are _______ and _______.
Q.8. The primary organ for accumulation (storage) of copper in the body is _______.
Q.9. The major route of elimination for copper from body is _______.
Q.10. Molybdenum and sulfur reduce the toxicity of copper by increasing _______.
Q.11. Anaemia caused by copper toxicosis is _______ type.
Q.12. Gunmetal kidneys are characteristic in _______ poisoning.
Q.13. In copper poisoning, elevation of liver-marker enzymes (AST, SDH, and LDH) is seen prior to _______ (about 3–6 weeks before).
Q.14. Salt poisoning is commonly associated with deprivation of _______.
Q.15. The most susceptible species for salt poisoning are _______ and _______.
Q.16. Salt causes toxicity due to the development of _______ or _______ in blood and CSF.
Q.17. The major symptoms evident in salt poisoning are _______.
Q.18. The prognosis in case of salt poisoning is _______.
Q.19. Trivalent arsenic compounds are less toxic than _______ arsenicals.
Q.20. Rice-water stool is a characteristic feature of _______ poisoning.

Answers (Exercise 7)

1. D-Penicillamine
2. Sheep
3. Chicken
4. Molybdenum
5. Bedlington terrier
6. Molybdenum
7. Transcuperin and ceruloplasmin
8. Liver
9. Biliary
10. Excretion
11. Haemolytic
12. Copper
13. Haemolytic crisis
14. Water
15. Poultry and pigs
16. Hyperosmolality or hypertonicity
17. Nervous symptoms
18. Grave
19. Pentavalent
20. Mercury

Exercise 8

Q.1. The metallic poison that is considered the 'king of poisons and poison of kings' is _______.
Q.2. The treatment of molybdenum poisoning involves administration of _______.
Q.3. The most toxic gaseous form of arsenic that is released during the charging of storage batteries are _______.
Q.4. Among the arsenicals, the least toxic form is _______.
Q.5. Drinking water containing more than _______ of arsenic is considered potentially toxic to large animals.
Q.6. The specific antidote for arsenic poisoning is _______.
Q.7. The water-soluble derivative that is considered superior to BAL is _______.
Q.8. The treatment of molybdenum poisoning involves administration of _______.
Q.9. Acute zinc deficiency causes _______ in children.
Q.10. The common source of mercury poisoning in animals is through _______.
Q.11. Sheep may be affected with arsenic toxicity due to the management practice of _______ for controlling ecto-parasites
Q.12. The species of animal that is more sensitive to arsenic poisoning is _______.
Q.13. Arsenic undergoes _______ biotransformation reaction in the body.
Q.14. The mechanism of toxicity of arsenic involves binding to _______.
Q.15. During acute arsenic toxicity, the primary symptom is _______.
Q.16. The nature of diarrhoea in acute arsenic poisoning is described as _______.
Q.17. In chronic arsenic poisoning, _______ coloured mucosa is characteristic.
Q.18. Samples that should be collected in suspected cases of arsenic toxicity are _______ and _______.
Q.19. The level of arsenic in visceral organs, indicative of arsenic poisoning, is _______.
Q.20. The diagnosis of oxalate crystals in _______ and _______ is indicative of oxalate poisoning.

Answers (Exercise 8)

1. Arsenic
2. Copper salts
3. Arsine
4. Organic arsenicals
5. 0.25% of arsenic
6. British anti-Lewisite (BAL) or dimercaprol
7. MDSA (Mesodimercapro-succinic acid), DMSA (di-mercapto-succinic acid)
8. Copper salts
9. Erosive dermatitis
10. Food
11. Dipping
12. Cat
13. Methylation
14. Sulphydryl groups (–SH)
15. Gastroenteritis
16. Rice watery
17. Brick red
18. Hair and nails
19. >3 ppm;
20. Kidney and rumen epithelium

Exercise 9

Q.1. The primary source for molybdenum in animals is through _______.
Q.2. The species of animal that is more susceptible to molybdenum poisoning is _______.

 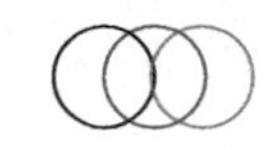

Q.3. Molybdenum toxicosis occurs primarily in the deficiency of _______ mineral intake.
Q.4. The ideal ratio of copper to molybdenum should be _______.
Q.5. The level of molybdenum that causes toxicity, irrespective of the copper content, is _______.
Q.6. Molybdenum is primarily excreted through _______.
Q.7. In molybdenum poisoning, deficiency of _______ mineral is observed.
Q.8. Peat scours or shooting diarrhoea is a characteristic symptom of _______.
Q.9. Light-coloured hair and depigmentation around eyes in molybdenum poisoning causes _______ appearance.
Q.10. Molybdenosis in sheep is manifested as _______ or _______.
Q.11. The anaemia caused by copper toxicosis is _______ type.
Q.12. Gunmetal kidneys are characteristic in _______ poisoning.
Q.13. The major symptoms of copper toxicosis are _______ and _______.
Q.14. In copper poisoning, elevation of liver-marker enzymes _______ is seen prior to _______ (about 3–6 weeks before).
Q.15. Lead inhibits 'heme' synthesis through the inhibition of the enzyme _______.
Q.16. In dogs, the predominant symptoms of lead poisoning are _______.
Q.17. The characteristic histological picture in tubular cells of kidney in lead poisoning is the presence of _______.
Q.18. Neurological symptoms accompanied by GIT symptoms possibly indicate _______ poisoning.
Q.19. The predominant manifestation of lead poisoning in cattle is _______.
Q.20. The level of sodium in plasma and CSF that is diagnostic of salt poisoning is _______.
Q.21. The oxides of metals tend to be _______, but the oxides of non-metals tend to be _______.
Q.22. A major site of toxic action for metals is interaction with _______.
Q.23. _______ is a low-molecular-weight metal-binding protein.
Q.24. _______ (name the important metal) is known to produce testicular injury after acute exposure.
Q.25. Cadmium can be found in the circulatory system bound to _______ and form a complex _______.
Q.26. Metallic arsenic is not _______ because it is insoluble in water and cannot be absorbed from the GI tract.
Q.27. The term 'itai-itai disease' was coined by locals for the severe pains felt in the _______ and _______ of victims.
Q.28. _______ is the specific antidote for iron toxicity.
Q.29. Arsenic is rapidly _______ and is completely eliminated in few days.
Q.30. The half-life of cadmium is about _______.

Answers (Exercise 9)

1. Grazing
2. Cattle
3. Copper
4. 6:1
5. >10 ppm
6. Urine
7. Copper
8. Molybdenosis
9. Spectacle-eye
10. Enzootic ataxia or sway back
11. Haemolytic

12. Copper
13. Hepatotoxicity and haemolytic anaemia
14. AST, SDH, and LDH; haemolytic crisis
15. δ-amino levulinic acid synthase (ALA-D synthase)
16. Gastrointestinal
17. Intranuclear inclusion bodies (eosinophilic)
18. Lead
19. Neurological symptoms.
20. >1 mEq/L
21. Basic, acidic
22. Enzymes
23. Metallothionein
24. Cadmium
25. Metallothionein (MT), CdMT
26. Poisonous
27. Spine, joints
28. Desferrioxamine (Desferrioxamine)
29. Detoxified
30. 30 years

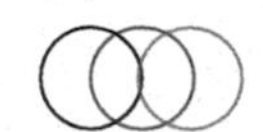

7.4 MATCH THE STATEMENTS

(Column A with Column B)

Exercise 10

Column A			Column B
Q.1.	Burton line	a.	Minamata disease
Q.2.	methylmercury	b.	arsenic
Q.3.	raindrop pigmentation	c.	lead
Q.4.	metallothionein	d.	overgrown hooves
Q.5.	selenium	e.	phosphorus
Q.6.	aluminum phosphide	f.	cadmium
Q.7.	sodium thiosulfate solution	g.	sulfur
Q.8.	beauty mineral	h.	iodine
Q.9.	teeth and bones	i.	copper
Q.10.	molybdenum	j.	fluoride

Answers (Exercise 10)

1. c 2. a 3. b 4. f 5. d 6. e 7. h 8. g 9. j 10. i

CHAPTER 8

TOXIC EFFECTS OF NON-METALS AND MICRONUTRIENTS

8.1 MULTIPLE-CHOICE QUESTIONS

(Choose the most appropriate response. It may be one, two, or more or none.)

Exercise 1

Q.1. Which of the following form of phosphorus is toxic?
- a) White
- b) Red
- c) Yellow
- d) Black

Q.2. Fluoride inhibits pyruvic-acid synthesis by inhibiting an enzyme ______.
- a) enolase
- b) transaminase
- c) phosphatase
- d) phosphodiesterase

Q.3. The most serious consequence of crude oil or kerosene ingestion by cattle is ______.
- a) liver damage
- b) kidney damage
- c) aspiration pneumonia
- d) central nervous system stimulation
- e) leukaemia

Q.4. After a period of drought or cloudy weather, the cyanide content in the plant ______.
- a) increases
- b) decreases
- c) does not change
- d) becomes zero

Q.5. Cyanide has more affinity for the following:
- a) Haemoglobin
- b) Cytochrome oxidase
- c) Met-haemoglobin
- d) Myoglobin

Q.6. The oxalate salt of the following element(s) is/are soluble:
- a) Sodium
- b) Potassium
- c) Magnesium
- d) Calcium

Q.7. The most appropriate sample to be collected in case of nitrate/nitrite poisoning is ______.
- a) blood
- b) rumen contents
- c) fodder
- d) CSF
- e) aqueous humour
- f) all of above

Q.8. Which of the following leads to the deposition of oxalate crystals in the urinary tract?
a) Dieffenbachia (dumbcane)
b) Propylene glycol
c) Methoxyflurane
d) Philodendron
e) Ethylene glycol

Q.9. Which of the following conditions represents a manifestation of moderate chronic fluoride toxicity?
a) Osteomalacia
b) Osteosclerosis
c) Osteopetrosis
d) Osteopenia
e) Osteolysis

Q.10. Intoxication from consumption of wild cherry or apricot pits would best be treated by ______.
a) hyperbaric oxygen
b) artificial respiration
c) inhalation of amyl nitrite
d) intravenous sodium nitrite and sodium thiosulfate
e) oral sodium nitrate

Answers (Exercise 1)

1. a & c 2. a 3. c 4. a 5. c 6. a & b 7. d 8. a 9. b 10. d

8.2 TRUE OR FALSE STATEMENTS

(Write T for True or F for False.)

Exercise 2

Q.1. In selenium toxicity, depleted glutathione levels can be restored by administering acetyl cysteine during treatment.
Q.2. Bran disease in horses is caused due to excess feeding of wheat bran, which contains high amount of phosphorus.
Q.3. In selenium poisoning, the level of selenium detected in the blood is 1–4 ppm and in hooves 5–20 ppm.
Q.4. The garlic-like odour of the breath or stomach contents luminous in the dark is indicative of selenium poisoning.
Q.5. Phosphorus is eliminated from the body through breath.
Q.6. Dermal exposure to white/yellow phosphorus leads to skin burns.
Q.7. The use of dimercaprol (BAL) metal chelator is contraindicated in selenium toxicity.
Q.8. The immediate symptom upon oral ingestion of phosphorus is emesis (hematemesis).
Q.9. The organs that are damaged in phosphorus poisoning are lungs and brain.
Q.10. In chronic phosphorus poisoning, the necrosis of the jaw that is observed is called itai-itai disease.
Q.11. The best material for the diagnosis of phosphorus poisoning is blood or sputim.
Q.12. The prognosis in case of phosphorus poisoning is very good after administration of chelating agents.
Q.13. Cyanide inhibits cellular respiration by binding with cytochrome oxidase.
Q.14. The conversion of methemoglobin to haemoglobin by methylene blue depends on the availability of oxygen.
Q.15. Oxalates belong to a group of substances known as nutrients.

Answers (Exercise 2)

1. T 2. T 3. T 4. F 5. T 6. T 7. T 8. T 9. F 10. F
11. F 12. F 13. T 14. F 15. F

8.3 MATCH THE STATEMENTS

(Column A with Column B)

Exercise 3

Column A		Column B	
Q.1.	bromine	a.	dull appearance
Q.2.	cyanide	b.	protoplasmic poison
Q.3.	halogeton glomeratus	c.	halogen
Q.4.	phosphorus	d.	reddish-brown liquid
Q.5.	non-metals	e.	goiter
Q.6.	oxalates	f.	skeleton
Q.7.	iodine	g.	histotoxic hypoxia
Q.8.	fluoride toxicity	h.	anti-nutrients
Q.9.	fluorine	i.	plant
Q.10.	chlorine	j.	osteosclerosis

Answers (Exercise 3)

1. d 2. g 3. i 4. b 5. a 6. h 7. e 8. j 9. f 10. c

Chapter 9

TOXICOLOGICAL HAZARDS OF SOLVENTS, GASES, VAPOURS, AND OTHER CHEMICALS

9.1 MULTIPLE-CHOICE QUESTIONS

(Choose the most appropriate response.)

Exercise 1

Q.1. Which of the following statements regarding solvents is *false*?
- a) Solvents can be absorbed from the GI tract and through the skin.
- b) Equilibration of absorbed solvents/vapours occurs most quickly in the lungs.
- c) Solvents are small molecules that lack charge.
- d) Volatility of solvents increases with molecular weight.
- e) Most solvents are refined from petroleum.

Q.2. What is the route in which most solvents enter the environment?
- a) Chemical spills
- b) Contamination of drinking water
- c) Evaporation
- d) Improper waste disposal
- e) Wind

Q.3. All of the following statements are true *except* ______.
- a) Most solvents can pass freely through membranes by diffusion.
- b) A solvent's lipophilicity is important in determining its rate of dermal absorption.
- c) Hydrophilic solvents have a relatively low blood:air partition coefficient.
- d) Biotransformation of a lipophilic solvent can result in the production of a mutagenic compound.
- e) Hepatic first-pass metabolism determines the amount of solvent absorbed in the GI tract.

Q.4. Which of the following statements regarding age solvent toxicity is ______?
- a) GI absorption is greater in adults than it is in children.
- b) Polar solvents reach higher blood levels in the elderly than they do in children.
- c) Children are always more susceptible to solvent toxicity than are adults.
- d) Increased alveolar ventilation increases the uptake of lipid-soluble solvents to a greater extent than water-soluble solvents.
- e) Increased body-fat percentage increases the clearance of solvent chemicals.

Q.5. Huffing gasoline can result in which of the following serious health problems?
- a) Renal failure
- b) Pneumothorax
- c) Hodgkin's disease
- d) Encephalopathy
- e) Thrombocytopenia

Q.6. Which of the following statements regarding benzene is *false*?
- a) A high-level exposure to benzene could result in acute myelogenous leukaemia (AML).
- b) Gasoline vapour emissions and auto exhaust are the two main contributors to benzene inhalation.
- c) Benzene is used as an ingredient in unleaded gasoline.

d) Benzene metabolites covalently bind DNA, RNA, and proteins and interfere with their normal functioning within the cell.
e) Reactive oxygen species can be derived from benzene.

Q.7. Which of the following is *not* a criterion for fetal alcohol syndrome diagnosis?
a) Maternal alcohol consumption during gestation
b) Pre and postnatal growth retardation
c) Microcephaly
d) Ocular toxicity
e) Mental retardation

Q.8. Which of the following is *not* an important enzyme in ethanol metabolism?
a) Alcohol dehydrogenase
b) Formaldehyde dehydrogenase
c) CYP2E1
d) Catalase
e) Acetaldehyde dehydrogenase

Q.9. Which of the following is *not* associated with glycol ether toxicity?
a) Irreversible spermatotoxicity
b) Craniofacial malformations
c) Hematotoxicity
d) Seminiferous tubule atrophy
e) Cleft lip

Q.10. Which of the following statements regarding chlorinated hydrocarbons is *false*?
a) Toxicities of trichloroethylene (TCE) are mediated mostly by reactive metabolites, not the parent compound.
b) Glutathione conjugation is an important metabolic step of both trichloroethylene (TCE) and perchloroethylene (PERC).
c) Many chlorinated hydrocarbons are used as degreasing agents.
d) Chloroform interferes with intracellular calcium homeostasis.
e) Carbon tetrachloride causes hepatocellular and kidney toxicity.

Q.11. The most serious consequence of crude oil or kerosene ingestion by cattle is ______.
a) liver damage
b) kidney damage
c) aspiration pneumonia
d) central nervous system stimulation
e) leukaemia

Q.12. In regard to chemically induced adverse effects on the eye ______.
a) no chemical has been shown to cause glaucoma
b) non-ionic detergents damage the eye more that cationic detergents
c) 2,4-dinitrophenol, corticosteroids, and naphthalene are known to cause cataracts in humans
d) methanol produces blindness by rendering the cornea and lens opaque
e) acids usually produce late-appearing ocular toxicity as contrasted to alkalis, which produce immediate damage

Q.13. Which of the following agents would *not* likely produce reactive airways dysfunction syndrome (RADS)?
a) Carbon monoxide
b) Chlorine
c) Ammonia
d) Toluene diisocyanate
e) Acetic acid

 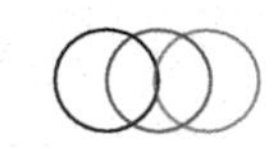

Q.14. Methyl bromide (CH_3Br) ______.
- a) is a liquid used primarily as a fumigant
- b) has essentially no warning properties, even at physiologically hazardous concentrations
- c) is extremely flammable
- d) is of greater concern from its oral toxicity than from its inhalation toxicity
- e) would not be expected to be readily absorbed through the lungs

Q.15. Benzene is similar to toluene ______.
- a) in its metabolism to redox active metabolites
- b) regarding covalent binding of its metabolites to proteins
- c) in its ability to produce CNS depression
- d) in its ability to produce acute myelogenous leukaemia
- e) in its ability to be metabolized to benzoquinone

Q.16. Sorbitol and other sugar alcohols have been associated with ______.
- a) respiratory distress syndrome
- b) osmotic diarrhoea
- c) hepatotoxicity
- d) immediate hypersensitivity reaction
- e) CNS depression

Q.17. Chloroform is *not* ______.
- a) central nervous system depressant
- b) hepatotoxic
- c) metabolized to phosgene
- d) peroxisome proliferator
- e) contaminant of chlorinated water

Q.18. Each of the following solvents is paired with a correct target organ of toxicity *except* ______.
- a) methanol:retina
- b) ethylene glycol:kidney
- c) ethylene glycol monomethyl ether:kidney
- d) dichloromethane:central nervous system
- e) carbon tetrachloride:liver

Q.19. Which of the following is *not* associated with spermatotoxicity in rats?
- a) Ethylene glycol monomethyl ether
- b) Ethylene glycol monoethyl ether
- c) Ethoxy acetic acid
- d) Methoxy acetic acid
- e) Propylene glycol monomethyl ether

Answers (Exercise 1)

1. d	2. c	3. c	4. b	5. d	6. b	7. d	8. b	9. a	10. d
11. c	12. c	13. a	14. b	15. c	16. b	17. d	18. c	19. e	

9.2 MATCH THE STATEMENTS

Exercise 2

(Column A with Column B)

Column A		Column B	
Q.1.	ethylene glycol	a.	hematopoietic toxicity
Q.2.	benzene	b.	hepatotoxicity
Q.3.	alkyl benzene	c.	reproductive toxicity
Q.4.	methanol	d.	pulmonary toxicity
Q.5.	ethanol	e.	CNS toxicity
Q.6.	chlorinated hydrocarbons	f.	ocular toxicity

Answers (Exercise 2)

1. c 2. a 3. e 4. f 5. b 6. d

CHAPTER 10

HAZARDS OF RADIATION AND RADIOACTIVE MATERIALS

10.1 MULTIPLE-CHOICE QUESTIONS

(Choose the most appropriate response.)

Exercise 1

Q.1. Which of the following ionizing radiations has the shortest range (i.e., travels the shortest distance in tissue) for the same initial energy?
   a) Alpha particle
   b) Beta particle
   c) Gamma ray
   d) X-ray
   e) Cosmic ray

Q.2. An individual exposed to 10 rads (0.1 Gy) of whole-body x-irradiation would be expected to ______.
   a) have a severe bone marrow depression
   b) die
   c) be permanently sterilized
   d) exhibit no symptoms
   e) vomit

Q.3. The cellular component that is affected during radiation damage is ______.
   a) lipid
   b) DNA
   c) RNA
   d) sugar

Q.4. Molecules with unpaired electrons in the outer shells are known as ______.
   a) free radicals
   b) sulfur
   c) nitrogen
   d) carbon

Q.5. Cells are more susceptible for radiation in the following stage(s) of the cell cycle:
   a) M-Phase (mitosis)
   b) Early G-Phase
   c) Late G-Phase
   d) S-phase

Q.6. The following organ is resistant to radiation:
   a) Endocrine glands
   b) Kidney
   c) Bone marrow
   d) Germinal cells

Q.7. More than 99% of the energy from the sun is within the spectral range of ______ to ______ nanometres.
   a) 150–4000
   b) 700–4000

c) 140–400
d) 400–700

Q.8. Which of the following inversions is short-lived?
a) Radiation
b) Subsidence

Q.9. The most common types of free radicals produced in the body are ______.
a) NO
b) O_2
c) SO_2
d) Reactive oxygen species (ROS)

Q.10. The part of the head that can be affected by the heating effects of a mobile phone is ______.
a) ears
b) nose
c) tongue
d) cornea

Q.11. Which of the following is *not* a main type of radiation?
a) Alpha particles
b) Microwaves
c) Beta particles
d) Gamma rays
e) X-rays

Q.12. Which of the following statements regarding alpha particles is *false*?
a) Alpha particles are ejected from the nucleus of an atom.
b) The atomic number decreases by two after emission of an alpha particle.
c) The atomic weight decreases by two after emission of an alpha particle.
d) Energies of most alpha particles range between 4 and 8 MeV.
e) Alpha particles are helium nuclei.

Q.13. Which of the following types of radiation is likely the *most* energetic?
a) Alpha particles
b) Beta particles
c) Positron emission
d) Electron capture
e) Photon emission

Q.14. Pair production and the Compton effect characterize which types of radiation interaction with matter?
a) Alpha particles
b) Beta particles
c) Positron emission
d) Electron capture
e) Photon emission

Q.15. Which of the following statements regarding radiation DNA damage is *false*?
a) Ionizing radiation slows down by forming ion pairs.
b) A main form of radiation DNA damage occurs by the production of free radicals.
c) High-LET radiation causes more ionizations than does low-LET radiation.
d) Most DNA damage caused by radiation happens directly.
e) Direct and indirect ionization cause similar damage to DNA.

Q.16. Low-LET radiation ______.
a) causes large-scale ionizations throughout the cell
b) results from alpha particle emission
c) causes damage that is readily repaired by cellular enzymes
d) is also known as densely ionizing radiation
e) usually causes irreparable cell damage

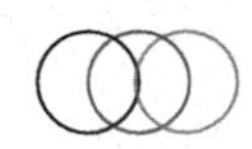

Q.17. What is the most common type of DNA damage caused by low-LET radiation exposure?
a) Base damage
b) DNA protein cross-links
c) Single-strand breaks
d) Double-strand breaks
e) Thymine-dimer formation

Q.18. Which of the following statements regarding radon exposure is *false*?
a) Miners are exposed to increased environmental radon levels.
b) Radon exposure has been linked to the development of lung cancer.
c) Smokers are at a higher risk from radon exposure.
d) Radon levels are relatively higher in urban areas than in rural areas.
e) The use of open frames indoors increases radon exposure.

Q.19. The largest dose of radiation is received from which of the following sources?
a) Inhalation
b) In body
c) Cosmic
d) Cosmogenic
e) Terrestrial

Q.20. The largest contributor to the effective dose of radiation in the US population is which of the following?
a) Nuclear medicine
b) Medical X-rays
c) Terrestrial
d) Internal
e) Radon

Q.21. What is the expected response of an individual exposed to a single absorbed dose of 10 rads (0.1 Gy) of whole-body X-irradiation?
a) Severe bone marrow depression
b) Permanent sterilization
c) No adverse response
d) Vomiting

Answers (Exercise 1)

1. a	2. d	3. b	4. a	5. a & b	6. a	7. a	8. a	9. d	10. d
11. b	12. c	13. a	14. e	15. d	16. c	17. c	18. d	19. a	20. e
21. c									

10.2 TRUE OR FALSE STATEMENTS

(Write T for True or F for False.)

Exercise 2

Q.1. The process of transmission of electromagnetic waves through a medium or vacuum is known as electron.
Q.2. Ionizing radiations produce ions by knocking out UV-A TYPE radiation from atoms.
Q.3. The most-sensitive hematological test for detecting radiation damage is absolute lymphocyte count.
Q.4. The ionizing radiation that has the least penetrating capacity is gamma (γ) rays.
Q.5. The ionizing radiation that has the highest penetrating capacity is alpha rays.
Q.6. The types of UV rays that can only reach the earth are UV-A type.
Q.7. Non-ionizing radiations include visible light, infrared rays, microwaves, and radio waves.
Q.8. The SI unit of radiation is a RAD (rad), and the depreciated unit in the CGS system is a GRAY (Gy).
Q.9. One Gray is equal to 100 rad. (or 1 rad = 0.01 Gy).
Q.10. The instrument that is used to measure radiation is called an electromagnetic counter.
Q.11. The most potent among the free radicals is HYDROXYL (OH-).
Q.12. The damaging effects of free radicals on protein is fragmentation and on DNA is protein oxidation.
Q.13. During a nuclear explosion, the ascent and subsequent descent of radioactive material either in the vicinity or away from the site of explosion is known as fall out.
Q.14. Among the radioactive material produced by nuclear explosions, the elements with biological significance are strontium (ST90) and cesium (CS137).
Q.15. The environmental hazard from nuclear power plants is due to the release of radioactive coolants.
Q.16. The group of symptoms that appear within 48 h of exposure to radiation are known as the prodromal syndrome.
Q.17. The most common type of cancer induced by radiation is leukaemia.
Q.18. In the case of grazing animals on fall-out pastures, the system that is primarily affected is the brain.
Q.19. The amount of radio frequency radiation absorbed by the human body while using a mobile phone is measured by specific absorption rate (SAR).
Q.20. The sensitivity of rapidly proliferating cells to radiation is less than other cells.
Q.21. The system in the body that is more susceptible to radiation exposure is the hemopoietic system.
Q.22. The lethal dose of radiation in humans is expressed as $LD_{50/60}$.

Answers (Exercise 2)

1. F	2. F	3. T	4. F	5. F	6. T	7. T	8. T	9. T	10. F	11. T
12. F	13. T	14. T	15. T	16. T	17. T	18. F	19. T	20. F	21. T	22. T

 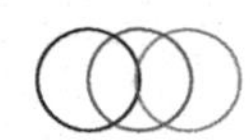

10.3 FILL IN THE BLANKS

Exercise 3

Q.1. During a nuclear explosion, the ascent and subsequent descent of radioactive material either in the vicinity or away from site of explosion is known as ______.

Q.2. Among the radioactive material produced by nuclear explosions, the elements with biological significance are ______ and ______.

Q.3. The environmental hazard from nuclear power plants is due to the release of radioactive ______.

Q.4. The group of symptoms that appear within 48 h of exposure to radiation is known as ______.

Q.5. The most common type of cancer induced by radiation is ______.

Q.6. In the case of grazing animals on fall-out pastures, the system that is primarily affected is ______.

Q.7. The amount of radio frequency radiation absorbed by the human body while using a mobile phone is measured by ______.

Q.8. The sensitivity of rapidly proliferating cells to radiation is ______.

Q.9. The system in the body that is more susceptible to radiation exposure is ______.

Q.10. The lethal dose of radiation in humans is expressed as ______.

Answers (Exercise 3)

1. Fall out
2. Strontium (ST90) and cesium (CS137)
3. Coolant
4. Prodromal syndrome
5. Leukaemia
6. Gastrointestinal system
7. Specific absorption rate (SAR)
8. More
9. Hemopoietic system
10. LD50/60

10.4 MATCH THE STATEMENTS

(Column A with Column B)-

Exercise 4

Column A		Column B	
Q.1.	37 billion	a.	heavy particles
Q.2.	energy in relation to radioactivity	b.	uranium decay products
Q.3.	alpha particles	c.	Curie
Q.4.	unpaired electrons	d.	endocrine glands
Q.5.	more susceptible for radiation	e.	reactive oxygen species
Q.6.	radon	f.	M-Phase (mitosis)
Q.7.	free radicals	g.	rad
Q.8.	resistant to radiation	h.	free electrons
Q.9.	Wilhelm Conrad Roentgen	i.	1945
Q.10.	Hiroshima	j.	X-ray

Answers (Exercise 4)

1. c 2. g 3. a 4. h 5. f 6. b 7. e 8. d 9. j 10. i

Chapter 11

TOXICITIES FROM HUMAN DRUGS

11.1 MULTIPLE-CHOICE QUESTIONS

(Choose the most appropriate response. It may be one, two, or more or none.)

Exercise 1

Q.1. Toxicity associated with any chemical substance is referred to as ______.
- a) poisoning
- b) intoxication
- c) overdose
- d) toxicology

Q.2. Clinical toxicity that is secondary to accidental exposure is ______.
- a) toxicology
- b) intoxication
- c) poisoning
- d) overdose

Q.3. Chest pain is related to ______.
- a) neurological examination
- b) cardiopulmonary examination
- c) GI examination
- d) both cardiopulmonary examination and GI examination

Q.4. A technique in which anticoagulated blood is passed through a column containing activated charcoal or resin particles is referred to as ______.
- a) whole-bowel irrigation
- b) forced dieresis
- c) haemodialysis
- d) haemoperfusion

Q.5. Which of the following substances is not easily adsorbed by activated charcoal?
- a) Iron
- b) Ethanol
- c) Methanol
- d) All of the above

Q.6. The effect of syrup of ipecac starts within 30 minutes of administration and lasts for approximately ______.
- a) 30 minutes
- b) 1 hour
- c) 1 hour and 30 minutes
- d) 2 hours

Q.7. Which of the following procedure(s) is/are contraindicated for patients who have ingested strong acids?
- a) Emesis
- b) Gastric lavage
- c) Whole-bowel irrigation
- d) Both emesis and gastric lavage

Q.8. Which technique is helpful in removing ethanol from the body?
- a) Dialysis
- b) Activated charcoal

c) Diuresis
d) Haemoperfusion

Q.9. The most effective treatment for GI decontamination with acetaminophen is ______.
a) emesis
b) gastric lavage
c) activated charcoal
d) dialysis

Q.10. Drug X is available as a 2.5% solution for intravenous administration. The desired dosage of this drug is 5 mg/kg. What volume of the drug should be injected if the animal weighs 50 kg?
a) 0.2 mL
b) 1.0 mL
c) 2.0 mL
d) 10 mL
e) 20 mL

Q.11. Thalidomide was accidentally discovered as a ______.
a) cardiotoxic agent
b) liver tonic
c) sedative/tranquilizer
d) cough mixture

Q.12. With barbiturate and BDZ abuse and dependency, sedative intoxication is generally associated with ______.
a) slurred speech
b) uncoordinated motor movements
c) impairment attention
d) all of the above

Q.13. A target organ of toxicity is ______.
a) lung
b) heart
c) reproductive system
d) kidney
e) liver

Q.14. A single large dose of *N*-nitrosodimethylamine fails to induce cancer in rats, but repeated dosing induces cancer because
a) the single large dose is lethal, while the threshold for cancer induction can be exceeded by repeated smaller doses
b) the main DNA lesion from a single large dose can be repaired readily by methyltransferase, while repeated smaller doses can deplete the available repair enzyme, induce mutations in DNA, and effectively induce cancer
c) the enzyme system involved in detoxification of *N*-nitrosodimethylamine is depleted after repeated doses, allowing *N*-nitrosodimethylamine to build up and exceed the threshold for cancer induction
d) the enzyme system involved in the conversion of *N*-nitrosodimethylamine to the active carcinogen is induced, and on subsequent repeated doses more active carcinogen is produced
e) the initial dose of *N*-nitrosodimethylamine causes cell damage and, thus, high mitotic rates; subsequent small doses induce mutations in DNA and effectively induce cancer

Answers (Exercise 1)

1. b	3. b	5. d	7. a and b	9. c	11. c	13. d
2. c	4. d	6. d	8. a	10. d	12. d	14. b

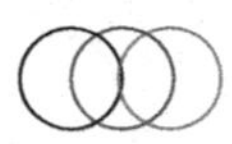

Exercise 2

Q.1. Which of the following drugs does *not* cause a prolonged QRS?
 a) Thioridazine
 b) Propranolol
 c) Quinine
 d) Metoprolol

Q.2. Which of the following antidotes is *not* used in cyanide poisoning?
 a) Dicobalt EDTA
 b) Hydroxocobalamin
 c) Sodium nitrite
 d) Dimercaprol

Q.3. Regarding "Tests for Drugs" in toxicology, which statement is *false*?
 a) Bedside electrocardiogram (ECG) and serum-paracetamol are regarded as routine toxicology screening tests.
 b) The fluorescence polarization immunoassay on urine or blood samples is used for "drug screening."
 c) Gas chromatography/mass spectrometry is performed as a "confirmatory test" on blood or urine samples.
 d) Thin layer/paper chromatography used on urine and blood samples assists in "drug screening."

Q.4. Extracorporeal elimination of drugs may be of use in all of the following *except* ______.
 a) ethylene glycol
 b) salicylates
 c) atenolol
 d) organophosphates

Q.5. The following statements about Digibind are true *except* ______.
 a) it is indicated when there is a history of ingestion of greater than 10 mg
 b) 40 mg binds approximately 0.6 mg digoxin
 c) serum digoxin levels increase following its administration
 d) indicated for use if serum digoxin level is greater than 10 nmol/L in acute overdose

Q.6. Which of the following is contraindicated to treat theophylline seizures?
 a) Diazepam
 b) Phenobarbitone
 c) Chloralhydrate
 d) Phenytoin

Q.7. Following an aspirin overdose, the initial acid-base derangement is usually ______.
 a) respiratory acidosis
 b) metabolic acidosis
 c) respiratory alkalosis
 d) metabolic alkalosis

Q.8. Which of the following pairs is *false* regarding drugs and their appropriate antidotes?
 a) β-Blockers – glucagon
 b) Chloroquine – diazepam
 c) Isoniazid – pralidoxime
 d) Methanol – ethanol

Q.9. With regard to sympathomimetic toxicity, which of the following is true?
 a) Co-ingestion of cocaine and alcohol results in greater neurological toxicity than cocaine alone.
 b) There is no difference between intravenous or oral amphetamine use and the incidence of rhabdomyolysis.

c) Patients with psychomotor acceleration and psychosis should initially be reviewed by the psychiatric team.
d) Auditory hallucinations are uncommon.

Q.10. The maximum safe dose for paracetamol every 24 hours is ______.
a) 90 mg/kg in children
b) 150 mg/kg in children
c) 200 mg/kg in children
d) in an adult up to 5 g

Q.11. Theophylline toxicity ______.
a) often presents with abdominal pain, hematemesis, and drowsiness
b) causes its effects by blockade of voltage-sensitive calcium channels in cardiac muscle and CNS
c) is rarely fatal with good supportive care
d) may cause refractory seizures

Q.12. Which statement is *false* regarding antihistamine toxicity?
a) Doxylamine overdose may result in non-traumatic rhabdomyolysis.
b) The first-generation antihistamine is a common cause for patients presenting with anticholinergic toxicity in ED.
c) Phenytoin is indicated for managing seizures.
d) Diphenhydramine and dimenhydrinate may cause cardiac conduction delays similar to tricyclic antidepressant (TCA) overdose.

Q.13. Which statement is *false* regarding colchicine poisoning?
a) Colchicine is rapidly absorbed following oral administration.
b) The multiorgan failure phase typically occurs 24 hours after ingestion.
c) A rebound leukocytosis occurs 3 weeks after poisoning in survivors, signaling recovery of bone marrow function.
d) Charcoal is indicated for gut decontamination.

Q.14. Which statement is true regarding anticonvulsant drug poisoning?
a) Chronic toxicity with therapeutic dosing is uncommon with phenytoin.
b) A poisoning with sodium valproate at 100 mg/kg is likely to result in a coma.
c) Cardiac monitoring is not required where phenytoin is the only agent ingested.
d) Carbamazepine levels are not useful in the management of carbamazepine poisoning.

Q.15. Regarding antimicrobial toxicity, the following are often fatal *except* ______.
a) isoniazid
b) neomycin
c) chloroquine
d) quinine

Q.16. Regarding isoniazid toxicity, all of the following are true *except* ______.
a) metabolic acidosis is common
b) treatment of seizures is best treated with high-dose BDZ
c) acidosis is thought to be secondary to seizures
d) toxicity is seen early post-ingestion (within 1–2 hours)

Q.17. Which is *false* regarding quinine poisoning?
a) Deliberate overdose is often fatal.
b) Significant overdose may result in cardiovascular collapse.
c) Deliberate overdose may result in permanent blindness.
d) PR interval prolongation is a major ECG change seen.

Q.18. In an overdose, the following are true *except* ______.
a) penicillins and cephalosporins are associated with seizures
b) seizures in isoniazid toxicity are due to pyridoxine disruption

 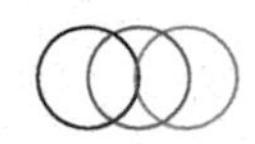

  c) a farmer presenting with hallucinations, dementia, and exquisitely painful legs may well have ergot poisoning
  d) a patient presenting after ingestion of colchicine presenting with minimal symptoms can be safely discharged after a short period of observation

Q.19. In a patient with a history of unknown psychiatric medication, which of the following is true?
  a) Hyperreflexia, rigidity, and hyperthermia would likely represent a dose-related effect of olanzapine.
  b) Extrapyramidal effects make it more likely to be a typical than atypical antipsychotic.
  c) Positive ECG changes in TCA toxicity are highly predictive of likely arrhythmias.
  d) If you suspect serotonin syndrome in an intubated patient, then midazolam and fentanyl sedation would be a good choice due to its short duration of action.

Q.20. Which of the following is *least* likely to be helpful in a calcium channel?
  a) Blocker overdose
  b) Atropine
  c) Intra-aortic balloon counter pulsation
  d) Insulin
  e) Resonium

Answers (Exercise 2)

1. d	2. d	3. d	4. d	5. d	6. d	7. c	8. c	9. d	10. a
11. d	12. c	13. c	14. c	15. b	16. b	17. d	18. d	19. b	20. d

11.2 FILL IN THE BLANKS

Exercise 3

Q.1. A target organ of toxicity is _______.
Q.2. What type of allergic contact dermatitis is _______.
Q.3. Duration of ultrashort-acting barbiturate is _______.
Q.4. Specific chemical used for cyanide poisoning is _______.
Q.5. _______ is contraindicated to treat theophylline seizures.

Answers (Exercise 3)

1. Kidney
2. Delayed type IV hypersensitivity reaction
3. 15–20 minutes
4. Sodium nitrite
5. Phenytoin

SECTION 5
PLANTS

CHAPTER 12

TOXIC EFFECTS OF PLANTS

12.1 MULTIPLE-CHOICE QUESTIONS

(Choose the most appropriate response. It may be one, two, or more or none.)

Exercise 1

Q.1. Consumption of milk from goats that have grazed on lupine plants containing the alkaloid anagyrine may cause ________.
  a) birth defects when ingested by women during early pregnancy
  b) severe liver damage characterized by centrolobular necrosis
  c) dizziness, nausea, headaches, and hallucinations
  d) numbness of the extremities
  e) aphrodisia and a general increase in sexual awareness

Q.2. During expression of oil from castor seeds, the toxic principle ricin is only present in ________.
  a) oil
  b) seed cake
  c) both
  d) none

Q.3. Intoxication from consumption of wild cherry or apricot pits would best be treated by ________.
  a) hyperbaric oxygen
  b) artificial respiration
  c) inhalation of amyl nitrite
  d) intravenous sodium nitrite and sodium thiosulfate
  e) oral sodium nitrate

Q.4. The imbalance of the following electrolyte is observed in oleander poisoning:
  a) Sodium
  b) Potassium
  c) Magnesium
  d) Calcium

Q.5. The contraindications in strychnine poisoning are ________.
  a) ketamine
  b) morphine
  c) emesis
  d) all

Q.6. The toxic part of the castor seed is the________.
  a) seeds
  b) oil
  c) stem
  d) leaves

Q.7. Active toxic principles of the croton plant is/are ________.
  a) calotoxin
  b) calotropin
  c) croton (toxalbumin)
  d) none of the above

Q.8. Which one is the toxic part of the *Calotropis* plant?
a) Stem
b) Branches
c) Milky white latex
d) All of the above

Q.9. The toxic principle of *Nerium odourum* is ________.
a) nerin
b) capsaicin
c) calotoxin
d) gigantin

Q.10. Which one is/are the toxic principle(s) of the aconite plant?
a) Aconitine
b) Misaconitine
c) Hypaconitine
d) All of the above

Q.11. Which of the following is responsible for the highest number of plant-caused fatalities in the western US?
a) Kochia weed
b) Sudan grass
c) Locoweed
d) Sorghum
e) Larkspur

Q.12. What is the difference between larkspur and monkshood?
a) Larkspur contains hollow stems and monkshood does not.
b) Monkshood contains hollow stems and larkspur does not.
c) Larkspur contains woody stems and monkshood does not.

Q.13. What is the treatment for larkspur?
a) Methylene blue
b) Acetyl cholinesterase inhibitor (physostigmine/neostigmine)
c) Sodium thiosulfate
d) All of the above

Q.14. A low-energy diet for ruminants may increase susceptibility to ________ poisoning?
a) Nitrite
b) Cardiac glycoside
c) Cyanide
d) Diterpenoid alkaloid
e) Phylloerythrin

Q.15. What is the characteristic clinical sign of cyanide poison?
a) Abortion
b) Circling
c) Cherry-red blood
d) Methaemoglobinemia
e) Brown mucous membranes

Q.16. What is the best method of treatment for nitrate poisoning?
a) Methylene blue given orally
b) Methylene blue given IM
c) Methylene blue given Sc
d) Methylene blue given IV
e) Methylene blue is not a treatment for nitrate poisoning

Q.17. Chronic Sudan grass poisoning of horse, sheep, and cattle causes ________.
a) demyelination
b) methaemoglobinemia

 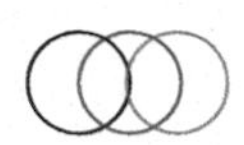

c) bad gas
d) CHF
e) cherry-red blood

Q.18. What plant contains a yellow fluid on cut surfaces that causes violent convulsions by blocking inhibitory pathways to brain?
a) Water hemlock
b) Kochia weed
c) Locoweed
d) Monkshood
e) Larkspur

Q.19. Sorghums and Sudan grasses are associated with which plant poison?
a) Nitrate
b) Nitrite
c) Cardiac glycoside
d) Diterpenoid alkaloids
e) Cyanide (prussic acid)

Q.20. Nitrate causes sudden death in cattle. Nitrate levels are highest on early morning/cloudy days due to the low nitrogen reductase levels. Which one of these plants has the highest level of nitrate?
a) Legumes
b) Corn
c) Sorghums
d) Annual weeds
e) Cereal grains

Q.21. This plant looks similar to an onion and causes excessive salivation, frothing, and sudden death in ewes:
a) Water hemlock
b) Monkshood
c) Larkspur
d) Death camas

Q.22. Which of the following are not considered cardiotoxic?
a) Yew
b) Foxglove
c) Avocado
d) Tansy ragwort

Q.23. What is the treatment for yew (taxine/taxane) poison?
a) Atropine sulfate
b) Methylene blue
c) Sodium thiosulfate
d) Charcoal
e) There is no treatment.

Q.24. What is the mechanism of action of cardiac glycosides?
a) Inhibit Na/K ATPase
b) Inhibit cytochrome oxidase
c) Inhibit acetylcholinesterase
d) Inhibit protein synthesis
e) Block ACh receptors

Q.25. What plant is associated with grayanotoxins that inhibits voltage-gated Na channels?
a) Locoweed
b) Milkweed
c) Oleander
d) Foxglove
e) Rhododendron

Q.26. St John's Wort and buckwheat are associated with ________.

a) primary photosensitization
b) secondary photosensitization

Q.27. Which organ is MOST severely damaged by pyrrolizidine alkaloids?

a) Heart
b) Spleen
c) Liver
d) Lung
e) Brain

Q.28. Which animal is most susceptible to plant alkaloids toxicity?

a) Horse
b) Cow
c) Chicken
d) Goat
e) Pig

Q.29. Which of the following plants is NOT associated with secondary photosensitization?

a) Hound's tongue
b) St John's Wort
c) Fiddle neck
d) Tansy ragwort
e) Rattle pod

Q.30. Which of the following plants is associated with selenium toxicity 'alkali disease'?

a) Milk vetch
b) Black walnut
c) Hoary alyssum
d) Sickle pod
e) Day-blooming jessamine

Q.31. Lectins are found in castor beans (ricin) and rosary peas. What is the mechanism of action of lectins?

a) Inhibit Na/K ATPase
b) Inhibit Na channels
c) Inhibit ACh receptors
d) Inhibit protein synthesis
e) Inhibit cell enzyme oxidation

Q.32. This plant causes more economic losses to the livestock industry than any other plant combined:

a) Castor beans
b) Rosary peas
c) Locoweed
d) Sorghums
e) Fox glove

Q.33. Yellow star thistle and Russian knapweed cause which disease?

a) Sleepy grass disease
b) Chewing disease
c) Alkali disease
d) Heart disease
e) Liver disease

Q.34. How do you treat bracken-fern toxicity?

a) There is no treatment
b) Acetylcholinesterase inhibitors
c) Sulfur-IV
d) Methylene blue-IV
e) Thiamine-IV

 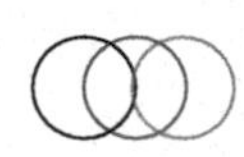

Q.35. Nightshade contains solanine toxin (muscarinic ACh receptor antagonism). What clinical sign is associated with this toxin?
a) Liver necrosis
b) First- and second-degree heart block
c) Urinary incontinence
d) Hypertension
e) Dilated pupils

Q.36. All of the following statements regarding plant toxicity are true *except*
a) Genetic variability plays a role in the toxicity of a plant.
b) Plant toxins are most highly concentrated in the leaves.
c) Young plants may have a higher toxin concentration than older plants.
d) The weather can influence the toxicity of plants.
e) Soil composition can alter a plant's production of toxin.

Q.37. Contact with which of the following plant species would be *least* likely to cause an allergic dermatitis?
a) *Urtica*
b) *Philodendron*
c) *Rhus*
d) *Dendranthema*
e) *Hevea*

Q.38. Which of the following statements regarding lectin toxicity is *false*?
a) Lectins have an affinity for *N*-acetyl glucosamine on mammalian neurons.
b) Consumption of lectins can cause severe gastrointestinal disturbances.
c) The fatality rate after ingestion of a fatal dose is very high.
d) Some toxic lectins inhibit protein synthesis.
e) A diet high in some lectins has been linked to reduced weight gain.

Q.39. Colchicine, found in lily bulbs, ________.
a) causes severe dehydration
b) is sometimes used as a purgative
c) causes a severe contact dermatitis
d) inhibits sphingolipid synthesis
e) blocks microtubule formation

Q.40. Activation of a vanilloid receptor is characteristic of which of the following chemicals?
a) Acetylandromedol
b) Capsaicin
c) Colchicine
d) Ergotamine
e) Linamarin

Q.41. Which of the following plant species is known to cause cardiac arrhythmias on ingestion?
a) *Dieffenbachia*
b) *Phytolacca americana*
c) *Digitalis purpurea*
d) *Pteridium aquilinum*
e) *Cicuta maculate*

Q.42. Which of the following plant toxins does *not* affect the neuromuscular junction?
a) Nicotine
b) Anabasine
c) Curare
d) Anatoxin A
e) Muscimol

Q.43. Which clinicopathologic value is least likely to be abnormal during 2 to 24 hours after a dog is bitten by a rattlesnake?
a) Serum amylase activity
b) Serum creatine phosphokinase activity
c) Serum-γ-glutamyltransferase activity
d) Platelet count
e) Prothrombin time

Q.44. Selenium absorption by crop plants is favoured by soil that is ________.
a) acidic, wet, and poorly drained
b) acidic, semi-arid, and well-drained
c) alkaline, well-aerated, and well-drained
d) alkaline, wet, and poorly aerated
e) acidic or alkaline, wet and poorly aerated

Q.45. In ruminants, urea toxicosis is characterized by ________.
a) ruminal alkalosis, systemic acidosis, and elevated blood ammonia levels
b) ruminal acidosis, systemic alkalosis, and elevated blood ammonia levels
c) ruminal alkalosis, systemic alkalosis, and elevated blood ammonia levels
d) ruminal acidosis, systemic alkalosis, and decreased blood ammonia levels
e) ruminal alkalosis, systemic alkalosis, and elevated blood urea nitrogen levels

Q.46. After a period of drought or cloudy weather, the cyanide content in the plant ________.
a) increases
b) decreases
c) does not change
d) becomes zero

Q.47. Cyanide has more affinity for the following
a) haemoglobin
b) cytochrome oxidase
c) met-haemoglobin
d) myoglobin

Q.48. The oxalate salt of the following element(s) is/are soluble:
a) Sodium
b) Potassium
c) Magnesium
d) Calcium

Q.49. The most appropriate sample to be collected in case of nitrate/nitrite poisoning is ________.
a) blood
b) rumen contents
c) fodder
d) CSF
e) aqueous humor
f) all of the above

Q.50. Which of the following leads to deposition of oxalate crystals in the urinary tract?
a) Dieffenbachia (dumbcane)
b) Propylene glycol
c) Methoxyflurane
d) Philodendron
e) Ethylene glycol

Q.51. Intoxication from consumption of wild cherry or apricot pits would best be treated by ________.
a) hyperbaric oxygen
b) artificial respiration

 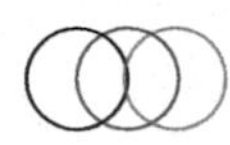

   c) inhalation of amyl nitrite
   d) intravenous sodium nitrite and sodium thiosulfate
   e) oral sodium nitrate

Q.52. The major toxic effect of hydrogen cyanide exposure is ________.
   a) lung damage
   b) haemoglobin alteration
   c) haemolysis of RBCs
   d) inhibition of mitochondrial respiration
   e) lipid peroxidation

Q.53. The parts of the plant that do not accumulate fluoride are the ________.
   a) seeds
   b) stems
   c) leaves
   d) flowers

Q.54. The following plant(s) is/are nitrate accumulators
   a) cereal grasses
   b) maize
   c) sunflower
   d) sorghum

Q.55. Which method(s) of storing forage reduces nitrate content?
   a) Hay making
   b) Silage making
   c) Composting
   d) Straw making

Q.56. Maximum accumulation of nitrates is seen in the following part of the plant ________.
   a) tip
   b) leaves
   c) upper 1/3 of stem
   d) lower 1/3 of stem

Q.57. During extraction of oil from castor seeds, the toxic principle ricin is only present in ________.
   a) oil
   b) seed cake
   c) both
   d) none

Q.58. Which of the following part of the bracken fern is more toxic?
   a) Rhizome
   b) Stem
   c) Leaves
   d) Tips

Q.59. Abrin is a ________.
   a) toxalbumin
   b) toxglobulin
   c) polypeptide
   d) carbohydrate

Q.60. The imbalance of the following electrolyte is observed in oleander poisoning ________.
   a) sodium
   b) potassium
   c) magnesium
   d) calcium

Q.61. The contraindications in strychnine poisoning are ________.
   a) ketamine
   b) morphine

c) emesis
d) all

Q.62. Convulsion in acute poisoning can be treated by ________.
a) morphine
b) barbiturates
c) d-tubocurarine
d) chloral hydrate

Q.63. Vitamin K is recommended in the treatment of poisoning due to ________.
a) sweet clover
b) zineb
c) prunus species
d) lotus species

Q.64. Liver damage occurs in poisoning with ________.
a) lantana
b) pyrrolizidine alkaloids
c) copper
d) all of the above
e) none of the above

Q.65. A garlic-like odour of the rumen content is indicative of ________.
a) phenol poisoning
b) phosphorus poisoning
c) hemlock poisoning
d) cyanide poisoning

Q.66. Cyanotic mucous membranes are seen in poisoning with ________.
a) nitrate/nitrite
b) chlorates
c) carbon monoxide
d) only one of the above
e) none of the above
f) all of the above

Q.67. The most appropriate sample to be collected in the case of nitrate/nitrite poisoning is ________.
a) blood
b) rumen contents
c) fodder
d) CSF
e) aqueous humor
f) all of the above

Answers (Exercise 1)

1. a	2. b	3. d	4. b	5. d	6. a	7. c	8. d	9. a	10. d
11. e	12. c	13. b	14. a	15. c	16. d	17. a	18. a	19. e	20. d
21. d	22. d	23. e	24. a	25. e	26. a	27. c	28. e	29. b	30. a
31. d	32. c	33. b	34. e	35. e	36. b	37. a	38. c	39. e	40. b
41. c	42. e	43. a	44. c	45. a	46. a	47. c	48. a & b	49. d	50. a
51. d	52. d	53. a	54. a, b, c, d	55. b	56. d	57. a	58. a	59. a	60. b
61. d	62. b	63. a	64. d	65. b	66. f	67. d			

 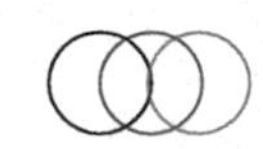

Exercise 2

Q.1. Which of the following plants are carcinogenic?
- a) American hellebore
- b) Bracken fern
- c) Buttercup
- d) Tung nut

Q.2. All of the following are true of grayanotoxins *except*
- a) They are present in rhododendron.
- b) They can contaminate honey.
- c) They cause bradycardia.
- d) They block the neuromuscular junction.

Q.3. Plant molecules that react as bases and usually contain nitrogen in a heterocyclic structure are known as ________.
- a) alkaloids
- b) terpenes
- c) resins
- d) glycosides

Q.4. Plant molecules that are created from isoprene units with varying functional groups are known as ________.
- a) terpenes
- b) amines
- c) alkaloids
- d) phenols

Q.5. Plant molecules that are hydrolysed to a sugar and a non-sugar moiety are known as ________.
- a) glycosides
- b) alkaloids
- c) resins
- d) terpene

Q.6. Toxic minerals that may accumulate in plants include all of the following *except* ________.
- a) cadmium
- b) magnesium
- c) copper
- d) selenium

Q.7. *Brassica oleracea* (kale) contains ________.
- a) cardiac glycoside
- b) cyanogenic glycoside
- c) goitrogenic glycoside
- d) steroid glycoside

Q.8. A plant toxin that can be highly transmitted through milk is ________.
- a) oxalate
- b) cyanide
- c) nitrate
- d) tremetol

Q.9. An antidote is available for the toxin present in ________.
- a) azalea
- b) pigweed
- c) veratrum
- d) apple seeds

Q.10. Which of the following plant toxins is classified as an alcohol?
- a) Nicotine
- b) Ranunculus

c) Dogbane
d) Tremetol

Q.11. Amygdalin is found in the highest amount in the seeds of ________.
a) bitter almond
b) tomato
c) pear
d) plum

Q.12. Strychnine blocks ________.
a) glycine-gated chloride channel
b) glutamate receptors
c) GABA receptors
d) voltage-gated sodium channels

Q.13. The toxin found in a species of *Capsicum* has been known to be useful in the therapy of ________.
a) skin cancer
b) depression
c) chronic pain
d) decubitus ulcers

Q.14. All of the following are true *except* ________.
a) it was a South American arrow poison.
b) it is a neuromuscular blocking agent.
c) it can be used clinically.
d) it is CNS toxic.

Q.15. Swainsonine ________.
a) is present in *Vinca* species
b) is a glycoprotein
c) causes abortions in livestock
d) none of the above

Q.16. The veratrum and lupine alkaloids are ________.
a) teratogenic
b) components of marketed pharmaceuticals
c) used as insecticides
d) poisons that Socrates drank

Answers (Exercise 2)

1. b 2. d 3. a 4. a 5. a 6. b 7. c 8. d 9. d 10. d
11. a 12. a 13. c 14. d 15. c 16. a

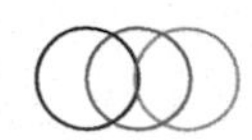

12.2 TRUE OR FALSE STATEMENTS

(Write T for True or F for False.)

Exercise 3

Q.1. The chronic selenium poisoning is also known as blind staggers.
Q.2. Selenium toxicity occurs when the pasture contains more than 5 ppm of selenium.
Q.3. The toxicity of nitrates is 6–10 times more than nitrite.
Q.4. Lantadin A&B from the *Lantana camera* causes secondary photosensitization.
Q.5. Aster sp. is an example of an indicator plant that contains a high concentration of selenium.
Q.6. Nitrate poisoning can cause abortion in cattle.
Q.7. There is a treatment for water hemlock (cicutoxin) poison.
Q.8. Enzootic hematuria due to bracken-fern poisoning is seen in a cattle species.
Q.9. Bracken fern produces permanent blindness in sheep species.
Q.10. The diagnosis of bracken-fern poisoning involves an estimation of the sugar level in blood.
Q.11. The specific treatment for bracken-fern-induced thiamine deficiency is vitamin D.
Q.12. The phytoconstituents of *L. camara* that cause bile-duct occlusion and liver damage are abrin ________.
Q.13. Hepatic lesions such as fibrosis and biliary hyperplasia are diagnostic in differentiation primary and secondary photosensitization.
Q.14. The primary visible lesion in photosensitization is spectacle eye.
Q.15. The prognosis is poor in primary-type photosensitization.

Answers (Exercise 3)

1. F 2. T 3. F 4. T 5. F 6. T 7. F 8. T 9. T 10. F
11. F 12. F 13. T 14. F 15. F

12.3 FILL IN THE BLANKS

Exercise 4

Q.1. *Abrus precatorius* is commonly known as ________ or ________.
Q.2. The toxicity due to seeds of *Abrus precatorius* is commonly referred to as ________ poisoning.
Q.3. The toxic principle present in the seeds of *Abrus precatorius* is ________.
Q.4. *Ricinus communis* is commonly referred to as ________.
Q.5. The toxic principle present in the seeds of *Ricinus communis* is ________.
Q.6. The toxic principles abrin and ricin belongs to the class of glycoproteins known as ________.
Q.7. The toxic principle present in the castor bean that causes haemagglutination and haemolysis is ________.
Q.8. The species of animal that is more susceptible to abrin and ricin poisoning is ________.
Q.9. At the cellular level, lectins (abrin and ricin) act by inhibiting ________ preventing protein synthesis.
Q.10. The toxic plant whose flowers are known as 'angel's trumpets' or 'moonflowers' is ________.
Q.11. The major tropane alkaloids present in *Datura stramonium* are ________ and ________.
Q.12. The toxic principles in *Dhatura* are present in ________ and ________.
Q.13. The plant that is known as the 'deadly night shade' is ________.
Q.14. The species of animal that is resistant to *Atropa belladonna* is ________.
Q.15. Tropane alkaloids (atropine) act by inhibiting ________ receptors in the body.
Q.16. The symptoms of Dhatura toxicity are termed as ________.
Q.17. Urine from *Dhatura*-intoxicated animals produces ________ in the cat, which is used for diagnosis.
Q.18. The competitive inhibitor of acetylcholine esterase that is used for treating *Dhatura* (atropine) intoxication is ________.
Q.19. Active toxic principles of *Abrus precatorius* is ________.
Q.20. Leftover cake after the extraction of oil is also ________.

Answers (Exercise 4)

1. Rosary pea or rathi
2. Sui/needle
3. Abrin
4. Castor bean
5. Ricin
6. Lectins
7. RCA (*Ricinus communis* agglutinin)
8. Horse
9. Ribosomes
10. *Datura stramonium*
11. Hyoscine atropine
12. Seeds, flowers
13. *Atropa belladonna*
14. Rabbit
15. Muscarinic
16. Anticholinergic delirium
17. Mydriasis
18. Physostigmine
19. Abrin
20. Highly toxic

 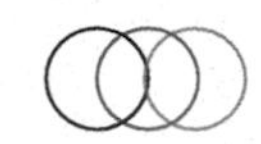

Exercise 5

Q.1. *Ipomoea turpethum* (Indian jalapa or morning glory) contains the toxic principle ________.
Q.2. The toxic principle present in *Ipomoea orizabensis* is ________.
Q.3. The main symptom in *Ipomoea* plant poisoning is ________.
Q.4. The toxic principles present in *Nerium oleander* (white oleander) is ________.
Q.5. The toxic principle present in *Nerium odorum* is ________.
Q.6. *Cerebra thevetia* contains the toxic glycosides ________ and ________.
Q.7. The steroidal glycoside present in *Nerium indicum* is ________.
Q.8. The most toxic part of the *Nerium* species of plants is ________.
Q.9. The species of animal most susceptible for oleander glycoside poisoning is ________.
Q.10. Oleander glycosides act by inhibiting ________ enzyme in cardiac cells.
Q.11. The toxic principle present in *Strychnos nuxvomica* is ________.
Q.12. The toxic principle strychnine is present in ________ part of *S. nuxvomica*.
Q.13. Strychnine produces spinal stimulation through inhibition of ________ neurotransmitter.
Q.14. The major site for action of strychnine in the spinal cord is ________ cells.
Q.15. In strychnine toxicity, the characteristic symptom is ________ appearance.
Q.16. The preferred sedative for strychnine poisoning is ________.
Q.17. The toxic principle present in cotton seeds is ________.
Q.18. Gossypol produces anaemia by inhibiting haeme synthesis by binding with ________.
Q.19. Inhibition of the testicular enzyme ________ is responsible for development of reproductive toxicity in males.
Q.20. Intake of high amounts of ________ is protective in gossypol toxicity.

Answers (Exercise 5)

1. Turpethum
2. Cammonin
3. Diarrhoea
4. Oleandrin
5. Nerin
6. Thevetin, cerebrin
7. Odoroside
8. Leaf
9. Horse
10. Na^+–K^+ ATPase
11. Strychnine
12. Seeds
13. Glycine
14. Renshaw
15. Saw horse
16. Pentobarbitone
17. Gossypol
18. Iron
19. Lactate dehydrogenase (LDH)
20. Protein

Exercise 6

Q.1. Oxalate poisoning in animals occurs commonly through ________ consumption.
Q.2. The most important oxalate-containing plants that cause oxalate poisoning are ________ and ________.
Q.3. *Halogeton* species contains ________% of oxalates on dry matter basis.
Q.4. The species of fungus that is rich in oxalates and can cause oxalate poisoning, is ________.

Q.5. The most commonly affected species for oxalate poisoning is ________.
Q.6. Pastures containing ________% oxalates is toxic for sheep.
Q.7. The important element with which oxalates interact in the body is ________.
Q.8. The primary clinical sign in oxalate poisoning is ________.
Q.9. The deposition of insoluble calcium oxalate crystals in renal tubules causes ________.
Q.10. The common site of urinary obstruction due to oxalate crystals in bulls and rams is ________.
Q.11. In oxalate poisoning, administration of ________ salts is carried out as a part of treatment.
Q.12. Nitrate is first converted to ________ in rumen by microflora.
Q.13. The use of ________ fertilisers increases the concentration of nitrates in plants.
Q.14. The herbicide that increases the nitrate content in plants is ________.
Q.15. High nitrate content is present in drinking water from ________ source.
Q.16. The species of animals that are more susceptible to nitrate poisoning are ________.
Q.17. The concentration of nitrates in plants that is toxic to animals is ________% or ________ ppm.
Q.18. Drinking water containing more than ________ ppm of nitrates can cause poisoning.
Q.19. The nitrate content in young plants is ________ than mature plants.
Q.20. The presence of ________ bacteria in water increases the nitrate toxicity.

Answers (Exercise 6)

1. Plants
2. *Halogeton glomeratus* and *Oxalis pes-caprae*
3. Q.34%
4. *Aspergillus*
5. Sheep
6. Q.2%
7. Calcium
8. Hypocalcaemia
9. Oxalate nephrosis
10. Sigmoid flexure
11. Calcium
12. Nitrite
13. Nitrate
14. 2, 4-D
15. Deep-well
16. Ruminants
17. 1% or 10,000 PPM
18. 1500 PPM
19. More
20. Coliform

Exercise 7

Q.1. The growth of plants/algal blooms in ponds causes ________ in the nitrate content of water.
Q.2. The deficiency of molybdenum/sulfur/phosphorus in soil causes ________ in nitrate accumulation in plants.
Q.3. The deficiency of copper/cobalt/manganese in soil causes ________ in nitrate accumulation in plants.
Q.4. The addition of ________ to feeds improves tolerance to nitrate toxicity.
Q.5. The antibiotic used as feed additive that enhances conversion of nitrates to nitrites causing poisoning is ________.
Q.6. Watering of animals immediately after consuming nitrate-rich plants ________ the chances of nitrate poisoning.

 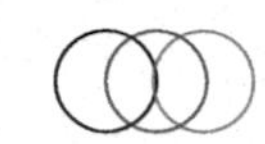

Q.7. Nitrite ions enter erythrocytes in exchange for ________ ion.
Q.8. Nitrite combines with haemoglobin (in a 1:2 ratio) to form ________.
Q.9. The colour of blood in nitrate poisoning is ________.
Q.10. Physiologically formed methaemoglobin (1 to 2%) is converted back to haemoglobin in the body by the enzymes ________.
Q.11. Formation of ________% of methaemoglobin produces symptoms of nitrate poisoning.
Q.12. Vasodilation and hypotension observed in nitrate poisoning is due to smooth-muscle relaxation caused by ________.
Q.13. The respiration in nitrate poisoning is ________.
Q.14. Chronic nitrate poisoning leads to the development of ________ in sheep.
Q.15. The preferred antemortem sample to be collected in nitrate poisoning is ________.
Q.16. The specific treatment for nitrate poisoning is ________.
Q.17. Methylene blue is oxidized to ________ during the conversion of methaemoglobin to haemoglobin.
Q.18. The conversion of methaemoglobin to haemoglobin by methylene blue depends on the availability of ________.
Q.19. The use of ________ is contraindicated in nitrate toxicity to improve blood pressure and heart function.
Q.20. The use of ________ metal chelates is contraindicated in selenium toxicity.

Answers (Exercise 7)

1. Decrease
2. Increase
3. Decrease
4. Soluble carbohydrates (TDN)
5. Monensin
6. Decreases
7. Chloride
8. Methaemoglobin
9. Chocolate brown
10. Diaphorase – I and II
11. 20% to 40%
12. Nitric oxide (NO)
13. Rapid
14. Goiter
15. Plasma
16. Methylene blue
17. Leucomethylene blue
18. NADPH2
19. Adrenergic agonists
20. Dimercaprol (BAL)

Exercise 8

Q.1. Cyanide from plant sources is present in the form of ________.
Q.2. The accumulation of ________ in plants predisposes to cyanide formation.
Q.3. The enzyme present in plants that releases hydrocyanic acid from cyanogenic glycosides is ________.
Q.4. The species of animals that are more susceptible to cyanogenic glycoside poisoning are ________.
Q.5. Among ruminants, the more susceptible species for cyanogenic glycosides is ________.
Q.6. The cyanogenic glycosides present in bitter almond and wild cherry is ________.

Q.7. The cyanogenic glycosides present in sorghum and Sudan grass is ________.
Q.8. The cyanogenic glycosides present in linseed and wild clover is ________.
Q.9. The part of the plant that contains more cyanogenic glycosides is ________.
Q.10. The level of HCN in plants that can cause cyanide poisoning in animals is ________.
Q.11. Young plants contain ________ cyanogenic glycosides than mature plants.
Q.12. The use of ________ fertilizers increases cyanide toxicity.
Q.13. Spraying of the weedicide ________ can increase cyanide content in plants.
Q.14. The soils that favour cyanide accumulation in plants are high in ________ and low in ________ content.
Q.15. Watering animals after feeding on cyanogenic plants ________ toxicity.
Q.16. The metabolite of cyanide that is excreted through urine is ________.
Q.17. Cyanide is converted to non-toxic thiocyanate by the enzyme ________.
Q.18. Good reserves of ________ in the body reduces the toxic effects of cyanide toxicity.
Q.19. Cyanide inhibits cellular respiration by binding with ________.
Q.20. The colour of blood in cyanide poisoning is ________.
Q.21. The characteristic smell of ruminal contents that can suggest cyanide poisoning is ________ smell.
Q.22. Chronic form of cyanide toxicity observed in humans due to consumption of cassava root is called ________.
Q.23. The level of cyanide in rumen contents that is indicative of cyanide poisoning is ________.
Q.24. The specific treatment for cyanide toxicity is ________ followed by ________.
Q.25. In selenium toxicity, depleted glutathione levels can be restored by administering ________ during treatment.
Q.26. In selenium poisoning, the level of selenium detected in blood is ________ ppm and in hooves is ________ ppm.
Q.27. In cattle, cracked and overgrown hooves are the main symptoms ________ poisoning.
Q.28. In horses, loss of hair from the mane is the main symptom of ________ poisoning.
Q.29. The important natural source of selenium apart from plants is ________.
Q.30. The maximum permissible amount of cotton seed cake that can be added to a ration of cattle is ________ kg. (total dose)

Answers (Exercise 8)

1. Cyanogenic glycosides
2. Nitrates
3. β-glycosidase
4. Ruminants
5. Cattle
6. Amygdaline
7. Dhurrin
8. Linamarine
9. Leaf
10. 200 PPM and above
11. More
12. Nitrate
13. 2, 4-D
14. Nitrogen, phosphorus
15. Increases
16. Thiocyanate
17. Rhodanese
18. Sulfur

19. Cytochrome oxidase
20. Bright red
21. Bitter almond
22. Konzo
23. 10 PPM
24. Sodium thiosulphate
25. Acetyl cysteine
26. 1–4 PPM, 5–20 PPM
27. Selenium
28. Selenium
29. Volcanic gases
30. One

Exercise 9

Q.1. The death in zinc phosphide poisoning is due to release of ________.

Q.2. The cherry-red colour of the mucous membrane is indicative of ________ poisoning.

Q.3. The deep-yellow colour of urine is seen in ________ poisoning, whereas a brown-black colour is indicative of ________ poisoning.

Q.4. The toxic principle of *Cannabis sativa* is ________.

Q.5. The cyanogenetic glycoside in *Sorghum* species is ________ and that in *Prunus* species is ________.

Q.6. ________ and ________ are used in the treatment of cyanide poisoning.

Q.7. Fagopyrin from *Fygopyrum esculantum* is responsible for causing ________ in cattle.

Q.8. Chronic copper poisoning in sheep occurs when the grazing pasture contains ________ ppm of copper on dry-matter basis.

Q.9. Ammonia exerts its toxic effect by inhibition of ________.

Q.10. Cyanide in smaller quantity is detoxified by an enzyme ________.

Q.11. ________ and ________ are mostly sent to the chemical examiner for confirming chronic arsenic poisoning.

Q.12. In cyanide poisoning ________ and ________ are the organs to be dispatched for chemical analysis.

Q.13. Liebermann's test is employed for confirming ________ poisoning.

Q.14. ________ and ________ cause anoxia by combining haemoglobin.

Q.15. The species of animal that is more susceptible to secondary photosensitization due to pyrrolizidine alkaloids is ________.

Q.16. *Lantana camara* causes ________ photosensitization.

Q.17. Abnormal sensitivity of unpigmented or less pigmented parts of skin to sunlight due to the presence of photodynamic substances in peripheral circulation is known as ________.

Q.18. The type of photosensitivity produced due to direct ingestion of photodynamic substances or metabolically activated agents is called ________ photosensitization.

Q.19. The photodynamic substance formed due to bacterial break down of chlorophyll, which is responsible for secondary photosensitization, is ________.

Q.20. Secondary photosensitization in *Lantana camara* is mainly due to ________.

Q.21. The phytoconstituents of *L. camara* that cause bile-duct occlusion and liver damage are ________.

Q.22. Hepatic lesions such as ________ and ________ are diagnostic in differentiation primary and secondary photosensitization.

Q.23. The primary visible lesion in photosensitization are ________.

Q.24. The prognosis is poor in ________ type photosensitization.

Answers (Exercise 9)

1. Phosphine gas
2. Cyanide
3. Picrate, Acorn
4. Tetrahydro cannabinol
5. Dhurrin, Prunasin
6. Sodium nitrite and sodium thiosulfate
7. Primary photosensitization
8. 15–20 ppm
9. Citric acid cycle
10. Rodonase
11. Horn and hair
12. Liver and lung
13. Carbolic acid (phenol)
14. Carbon monoxide and nitrite
15. Pig
16. Secondary
17. Photosensitization
18. Primary
19. Phylloerythrin
20. Bile duct occlusion
21. Lantadene A & B
22. Fibrosis, biliary hyperplasia
23. Sunburns
24. Secondary

Exercise 10

Q.1. 1-Gossypol-induced infertility is due to ________ effect on the ovaries.
Q.2. Bracken-fern poisoning is caused by the consumption of the plant ________.
Q.3. The enzyme present in the bracken fern that breaks down vitamin ________ is ________.
Q.4. Aplasia of bone marrow due to bracken fern toxicity is observed in ________ species of animals.
Q.5. The aplastic anaemia cause and carcinogenic factor present in the bracken fern is ________.
Q.6. In alkaline pH, ptaquiloside is converted in to an active carcinogenic form ________.
Q.7. The co-carcinogen present in the bracken fern that causes malignant tumours in the mouth, esophagus, and rumen along with the papilloma virus is ________.
Q.8. The most susceptible species for bracken-fern poisoning are ________ and ________.
Q.9. Anaemia is absent in bracken-fern poisoning in ________ species.
Q.10. The main symptom of bracken-fern poisoning in cattle is ________.
Q.11. Long-term consumption of bracken fern causes ________ tumours in cattle.
Q.12. Enzootic hematuria due to bracken-fern poisoning is seen in ________ species.
Q.13. Bracken fern produces permanent blindness in ________ species.
Q.14. The diagnosis of bracken-fern poisoning involves estimation of ________ levels in the blood.
Q.15. The specific treatment for bracken-fern-induced thiamine deficiency is ________.

Answers (Exercise 10)

1. Luteolytic
2. *Pteridium aquilinum*
3. B1, thiaminase

4. Ruminants
5. Ptaquiloside
6. Dienon
7. Quercetin
8. Horse, cattle
9. Horse
10. Aplastic anaemia
11. Urinary bladder
12. Cattle
13. Sheep
14. Thiamine
15. Thiamine (Vit B1)

12.4 MATCH THE STATEMENTS

(Column A with Column B)

Exercise 11

Column A		Column B	
Q.1.	apricot	a.	anticholinergic
Q.2.	water hemlock	b.	nausea, vomiting
Q.3.	lupine	c.	soluble oxalate
Q.4.	cactus	d.	seizures, tremors
Q.5.	poinsettia	e.	CNS cognitive
Q.6.	oleander	f.	photosensitivity-inducing
Q.7.	jimsonweed	g.	cyanide
Q.8.	St. John's wort	h.	contact irritant dermatitis
Q.9.	rhubarb	i.	hepatotoxic and teratogenic
Q.10.	marijuana	j.	cardiac arrhythmias

Answers (Exercise 11)

1. g 2. d 3. i 4. h 5. b 6. j 7. a 8. f 9. c 10. e

Section 6
POISONOUS ORGANISMS, FOOD AND FEED TOXICITY

Chapter 13

BIOTOXINS AND VENOMOUS ORGANISMS

13.1 MULTIPLE-CHOICE QUESTIONS

(Choose the most appropriate response.)

Exercise 1

Q.1. Which of the following statements regarding animal toxins is *false*?
- a) Animal venoms are strictly metabolized by the liver.
- b) The kidneys are responsible for the excretion of metabolized venom.
- c) Venoms can be absorbed by facilitated diffusion.
- d) Most venom fractions distribute unequally throughout the body.
- e) Venom receptor sites exhibit highly variable degrees of sensitivity.

Q.2. Scorpion venoms do *not* ________.
- a) affect potassium channels
- b) affect sodium channels
- c) affect chloride channels
- d) affect calcium channels
- e) affect initial depolarization of the action potential

Q.3. Which of the following statements regarding widow spiders is true?
- a) Widow spiders are exclusively found in tropical regions.
- b) Both male and female widow spiders bite and envenomate humans.
- c) The widow spider toxin decreases the calcium concentration in the synaptic terminal.
- d) Alpha-latrotoxin stimulates increased exocytosis from the nerve terminals.
- e) A severe alpha-latrotoxin envenomation can result in life-threatening hypotension.

Q.4. Which of the following diseases is *not* commonly caused by tick envenomation?
- a) Rocky Mountain spotted fever
- b) Lyme disease
- c) Q fever
- d) Ehrlichiosis
- e) Cat scratch fever

Q.5. Which of the following is *not* characteristic Lepidoptera envenomation?
- a) Increased prothrombin time
- b) Decreased fibrinogen levels
- c) Decreased partial thromboplastin time
- d) Increased risk of hemorrhaging
- e) Decreased plasminogen levels

Q.6. Which of the following animals has a venom containing histamine and mast-cell-degranulating peptide that is known for causing hypersensitivity reactions?
- a) Bees
- b) Ants
- c) Snakes
- d) Spiders
- e) Reduviidae

Q.7. Which of the following enzymes is *not* typically found in snake venoms?
- a) Hyaluronidase
- b) Lactate dehydrogenase

c) Collagenase
d) Phosphodiesterase
e) Histaminase

Q.8. Which of the following statements regarding snakes is *false*?
a) Inorganic anions are often found in snake venoms.
b) About 20% of snake species are venomous.
c) Snake venoms often interfere with blood coagulation mechanisms.
d) Proteolytic enzymes are common constituents of snake venoms.
e) Snakebite treatment is often specific for each type of envenomation.

Q.9. Which of the following is *false* regarding botulism toxin?
a) ACh release is blocked in the presynaptic neuron resulting in flaccid paralysis.
b) Botulism occurs via ingestion or wound contamination of spores or preformed toxin.
c) Preformed toxin sources are decaying carcasses.
d) For prevention, vaccination against *C. botulism* with toxoid can prevent clinical disease.
e) Clinical signs include 'sawhorse stance', muscle rigidity, erect ears, and a reluctance to eat due to 'locked jaw'.

Q.10. Cyanobacteria can cause hepatoxicity via ingestion of ________ and neurotoxicity via ingestion of ________.
a) microcystin/nodularin, anatoxin
b) anatoxin, microcystin/nodularin

Q.11. Blister beetles cause cathardin toxicosis, which occurs when livestock eat ________.
a) grain
b) alfalfa hay
c) dead chicken carcasses
d) locoweed
e) cardenolide

Q.12. Which of the following is *false* regarding Bufo toad toxicity?
a) Onset of clinical signs can be rapid, and death can occur within 15 minutes.
b) The cardiac glycosides bind and inhibit Na/K ATPase resulting in a depressed electrical conduction.
c) Mucous membranes appear pale and tacky.
d) Prognosis is good with most animals with early decontamination and appropriate symptomatic therapy. Majority of animals present with neurologic abnormalities, including convulsions, ataxia, nystagmus, stupor, and coma.

Q.13. Which is *false* regarding pit vipers?
a) Most bites are by copperheads.
b) Venom is delivered by the retractable fangs downward and stabbing forward.
c) Echinocytes greatly increase the likelihood that the victim has been envenomed.
d) The initial clinical sign is usually marked by tissue swelling.
e) Antivenin treatment (CroFab™) substantially increases the likelihood of survival.

Q.14. How would you identify a female black widow spider?
a) Red, yellow, or orange hourglass on ventral abdomen
b) Bite causes a 'bull's-eye' lesion and systemic deletion of clotting factors (VII, IX, and XII)
c) Within 30 min of bite, expanding area of wound will reach up to 15 cm and rupture with serious discharge
d) Female black widow spiders only build nests in barns housed by talking pigs named Wilbur
e) Nests will be built on ground level since black widows are poor climbers

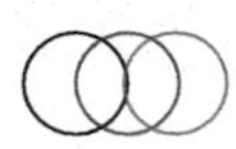

Q.15. Poisoning from these animals is referred to as tegenarism and can generate a wound that takes years to properly heal ________.

a) black widow
b) hobo spider
c) brown recluse
d) tarantula
e) scorpion

Q.16. Which species toxin forms a 'bull's-eye' lesion?

a) Black widow
b) Brown recluse
c) Hobo spider
d) Tarantula
e) Scorpion

Q.17. Antivenin (Lycovac) is available and effective for which species toxin?

a) Black widow
b) Brown recluse
c) Hobo spider
d) Tarantula
e) Scorpion

Answers (Exercise 1)

1. a	2. d	3. d	4. e	5. c	6. a	7. e	8. a	9. e	10. a
11. b	12. c	13. e	14. a	15. b	16. b	17. a			

13.2 TRUE OR FALSE STATEMENTS

(Write T for True and F for False.)

Exercise 2

Q.1. Cathardin toxicosis from blister beetles can cause irritation in the terminal ends of the esophagus, stomach, and intestines leading to ulcerative lesions.
Q.2. Coral snakes have short, fixed, non-hinged front fangs that are partially membrane covered, but 60% of bites do not result in venom delivery.
Q.3. Coral snake venom is rapidly cleared from the body.
Q.4. 99% of all the snakebites to animals in North America are by coral snakes.
Q.5. Tarantula species native to the United States are not capable of delivering serious envenomation and thus are effectively innocuous.
Q.6. Do not remove a tick; just provide respiratory support for treatment of tick toxin.

Answers (Exercise 2)

1. T 2. T 3. F 4. F 5. T 6. F

 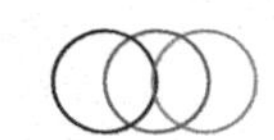

13.3 FILL IN THE BLANKS

Exercise 3

Q.1. The type of toxins present in the venom of elapidae snakes (cobra, krait, mamba, coral snakes) are ________.
Q.2. The type of toxins present in the venom of viperidae snakes (viper, rattlesnake, adder) are ________.
Q.3. α-Bungarotoxin is the neurotoxin present in the venom of ________ snakes.
Q.4. In snakes, the venom glands are homologous to ________ glands in other animals.
Q.5. The symptoms in elapine snake bites are predominantly ________.
Q.6. Diagnosis of a snake bite is possible by observing ________ marks on the body of the animal.
Q.7. The main treatment for snake bite is ________.
Q.8. The only poisonous lizard known to cause poisoning in animals and humans is ________.
Q.9. The inclusion of ________ component in feeds can expose animals to fish poisons.
Q.10. The most potent among all marine toxins is ________.
Q.11. The eating of ________ fish is commonly associated with tetrodotoxin poisoning.
Q.12. The organs in which tetrodotoxin (TTX) primarily accumulates in puffer fish are ________ and ________.
Q.13. Tetrodotoxin is produced by the bacteria ________, which live symbiotically with the puffer fish.
Q.14. Tetrodotoxin acts as a potent neurotoxin through blocking of ________ channel in the central and peripheral nervous system.
Q.15. Ciguatera is a fish-borne poisoning resulting from the consumption of ________ fish.
Q.16. The toxin responsible for ciguatera poisoning is ________.
Q.17. Ciguatoxin is produced by the dinoflagellate ________ present in reef fish.
Q.18. Consumption of tuna and mackerel fish is associated with ________ poisoning.
Q.19. Scombroid poisoning is caused by ________, which is produced from histidine by bacterial action.
Q.20. Local toxic effects are more prominent after the snake bite of the ________.

Answers (Exercise 3)

1. Neurotoxins
2. Hemotoxins
3. Krait
4. Parotoid
5. Neurological
6. Fang
7. Monovalent/polyvalent antivenin
8. *Heloderma suspectum* (gila monster)
9. Fish meal
10. Tetrodotoxin (TTX)
11. Tetraodon (puffer)/fugu
12. Liver and ovary
13. *Pseudoalteromonas tetradonis*
14. Sodium
15. Reef fish (barracuda, eel, etc.)
16. Ciguatoxin (CTX)
17. *Gambierdiscus toxicus*
18. Scombroid
19. Histamine
20. Rattlesnake

Exercise 4

Q.1. The order under the class Insecta that includes greatest number of poisonous insect species is ________.

Q.2. The antigenic component of honey bee venom that can cause allergies or anaphylactic shock is ________.

Q.3. The drug of choice for the treatment of systemic reactions produced by stinging from bees or wasps is ________.

Q.4. The potent cytotoxin present in ant venom is ________.

Q.5. The piperidine alkaloid component of fire ant venom is ________.

Q.6. Focal necrotic ulcers of the cornea and conjunctiva in calves are caused by the bite of the ________ insect.

Q.7. The most potent neurotoxin present in the venom of the black widow spider (*Lactrodectus mactans*) is ________.

Q.8. The most dangerous species of scorpion is ________.

Q.9. The most common tick species that are responsible for the development of tick paralysis are ________.

Q.10. Tick paralysis is caused by the injection of ________, which is neurotoxic.

Q.11. The most susceptible species for tick paralysis is ________.

Q.12. The type of paralysis seen in tick paralysis is ________ type.

Q.13. Dermacentor species toxin acts by blocking ________ channels, whereas Ixodidae ticks block the release of ________ in motor nerves.

Q.14. The diagnosis of tick poisoning is made by the detection of ________ sex ticks on the animal body along with symptoms of paralysis.

Q.15. The venom in toads is produced by ________ glands.

Q.16. The most toxic among the toads is ________.

Q.17. The principal component of toad venom is ________.

Q.18. The animal species that is more prone for toad poisoning is ________.

Q.19. Treatment of toad poisoning involves the use of ________ drug to control cardiac arrhythmia and fibrillation.

Q.20. A snake bite is commonly observed in ________ and ________ species.

Answers (Exercise 4)

1. Hymenoptera
2. Mellitin
3. Epinephrine (adrenaline)
4. Formic acid
5. Solenopsin
6. Fire ants
7. A-lacrotoxin
8. *Leiurus quinquestriatus*
9. *Rhiphicephalus* and dermacentor
10. Saliva
11. Dog
12. Ascending
13. Sodium, acetylcholine
14. Female
15. Parotoid
16. *Bufo marinus*
17. Bufodienolides (cardiac glycosides)
18. Dog
19. Propranolol
20. Dog, horse

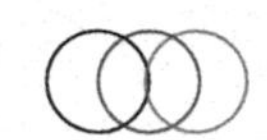

13.4 MATCH THE STATEMENTS

(Column A with Column B

Exercise 5

Column A		Column B	
Q.1.	*E. coli*	a.	mahi mahi
Q.2.	ciguatera poisoning	b.	GRAS substance
Q.3.	endotoxin	c.	gram-negative bacteria toxin
Q.4.	emetic toxin	d.	enzyme
Q.5.	fluoride	e.	apple products
Q.6.	scombroid poisoning	f.	beets
Q.7.	iron oxide	g.	*B. cereus*
Q.8.	rennet	h.	dinoflagellates
Q.9.	patulin	i.	contaminant in hamburger, meat, raw vegetables
Q.10.	high nitrates	j.	osteosclerosis

Answers (Exercise 5)

1. i 2. h 3. c 4. g 5. j 6. a 7. b 8. d 9. e 10. f

CHAPTER 14

FOOD HAZARDS AND FEED-CONTAMINANT TOXICITY

14.1 MULTIPLE-CHOICE QUESTIONS

(Choose the most appropriate response. It may be one, two, or more or none.)

Exercise 1

Q.1. Which of the following is *not* true regarding Amanita phalloides mushrooms?
- a) Toxic components are phalloidin and amatoxins
- b) Produces liver and gastrointestinal toxicity
- c) Cardiovascular toxicity is responsible for mortality
- d) Common name is 'death cap'
- e) No specific antidotal treatment of poisoning is available

Q.2. The following feed stuff supports growth of aflatoxins:
- a) Ground nut cake
- b) Soybean cake
- c) Cotton seed meal
- d) All

Q.3. Aflatoxin has the following character(s):
- a) Carcinogenic
- b) Mutagenic
- c) Teratogenic
- d) Immunosuppressive

Q.4. Which of the following forms of aflatoxicosis is most common?
- a) Per acute
- b) Acute
- c) Subacute
- d) Chronic

Q.5. Rubratoxins are destroyed at the following temperature ________.
- a) Freezing
- b) Room temperature
- c) 50°C–60°C
- d) 85°C–100°C

Q.6. The site of action of ochratoxins in the nephron is ________.
- a) proximal convoluted tubule
- b) loop of henle
- c) distal convoluted tubule
- d) collecting duct

Q.7. Which of the following serotype(s) of botulinum is most commonly implicated in animals and poultry?
- a) A-type
- b) B-type
- c) C-type
- d) D-type

Q.8. Which of the following toxicities are infectious?
a) Botulism
b) Tetanus

Q.9. The type of skeletal muscle contractions seen in tetanus are ________.
a) clonic
b) tonic
c) both
d) twitching

Q.10. What is the correct temperature at which frozen food should be kept?
a) 0°C
b) 15°C or lower
c) 8°C or lower
d) 20°C or lower

Q.11. Overheated Teflon-coated frying pans release vapours that are especially toxic to ________.
a) cats
b) dogs
c) gerbils
d) parakeets
e) reptiles

Q.12. A major mode of transmission of foodborne illnesses is ________.
a) Via mosquito transmission
b) Via fecal–oral route
c) Via person to person contact
d) Via hypodermic syringes

Q.13. Which of the following is/are economic consequences of foodborne illness?
a) Medical costs
b) Investigative costs
c) Loss of wages
d) Litigation costs
e) All of the above

Q.14. Of those listed below, which is *not* a foodborne pathogen?
a) Lectins
b) Nematodes
c) Bacteria
d) Protozoans

Q.15. Which of the following leads to neurologic complications?
a) Mercury
b) Lead
c) Cadmium
d) Antimony

Q.16. GRAS substances are those substances added to food that are ________.
a) generally responsible for acute sickness
b) government reported assumed safe
c) general response acidosis sickness
d) generally recognized as safe

Q.17. All of the following are gram-negative bacteria *except* ________.
a) *Staphylococcus aureus*
b) *Escherichia coli*
c) *Salmonella typhimurium*
d) *Vibrio cholerae*

 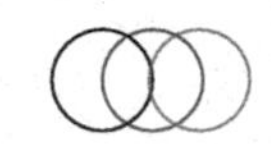

Q.18. Acidic conditions will leach all but which of the following from packaging material?
- a) Lead
- b) Cadmium
- c) Antimony
- d) Mercury

Q.19. *Taenia solium* (pork tapeworm) has more serious consequences than *Taenia saginata* (beef tapeworm) because it ________.
- a) has a hooked rostellum that attaches it to the intestine wall
- b) can migrate to the brain, eyes, and muscles
- c) can be ingested from the waste of another human
- d) all of the above

Q.20. Of those listed below, which is *not* a protozoan?
- a) *Cyrptosporidium*
- b) *Entamoeba histolytica*
- c) *Penicillium* spp.
- d) *Giardia lamblia*

Q.21. Which of the following statements regarding food complexity is *false*?
- a) Many flavour additives are non-nutrient substances.
- b) Foods are subjected to environmental forces that alter their chemical composition.
- c) There are more non-nutrient chemicals in food than nutrient chemicals.
- d) A majority of non-nutrient chemicals are added to food by humans.
- e) Food is more variable and complex than most other substances to which humans are exposed.

Q.22. Which of the following foods contains the most non-nutrient chemicals?
- a) Bees
- b) Banana
- c) Tomato
- d) Orange juice
- e) Cheddar cheese

Q.23. Which of the following is considered an indirect food additive?
- a) Nitrites
- b) Plastic
- c) Food colouring
- d) EDTA
- e) Citric acid

Q.24. Estimated daily intake (EDI) is based on which of the following?
- a) Metabolic rate
- b) Daily intake
- c) Substance concentration in a food item
- d) Body mass index
- e) Concentration of substance in a food item and daily intake

Q.25. Which of the following is *not* characteristic of IgE-mediated food allergies?
- a) Urticaria
- b) Wheezing
- c) Hypertension
- d) Nausea
- e) Shock

Q.26. Which of the following wheat proteins is famous for being allergenic?
- a) Casein
- b) Ovalbumin

c) Livetin
d) Gluten
e) Glycinin

Q.27. Which of the following foods contains a chemical that causes hypertension by acting as a noradrenergic stimulant?
a) Cheese
b) Peanuts
c) Shrimp
d) Chocolate
e) Beets

Q.28. What is the mechanism of saxitoxin found in shellfish?
a) Interference with ion channels
b) Direct neurotoxicity
c) Interference with DNA replication
d) Binding to haemoglobin
e) Interference with a stimulatory G protein

Q.29. Which of the following foods can cause a reaction that mimics iodine deficiency?
a) Chocolate
b) Shellfish
c) Peanuts
d) Fava beans
e) Cabbage

Q.30. Improperly canned foods can be contaminated with which of the following bacteria, causing respiratory paralysis?
a) *C. perfringens*
b) *R. ricketsii*
c) *S. aureus*
d) *C. botulinum*
e) *E. coli*

Q.31. *Hepatitis A* can be transferred via ________.
a) the fecal oral route
b) shellfish from polluted water
c) intravenous drug users
d) food handling by infected workers
e) all of the above

Q.32. Heterocyclic amines and acrylamide are food contaminants that ________.
a) are produced by microorganisms
b) are produced by the process of cooking
c) are considered as GRAS
d) are residues from animal feeds

Q.33. The concepts of 'de minimis' as applied to food safety means ________.
a) find the smallest harmful dose
b) only food colours 1/100 of the NOAEL can be used
c) pesticide residues can be present at the ADI
d) the risk is so small it is of no concern

Q.34. The primary reason for adding nitrates and nitrites to food is to ________.
a) prevent the growth of *Clostridium botulinum*
b) give the meat a characteristic flavour
c) turn the meat a brown-red colour
d) sweeten the food product

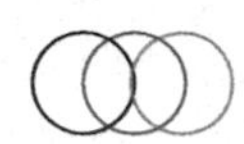

Q.35. Where should raw meat be stored in a refrigerator?
- a) At the top
- b) In the middle
- c) At the bottom, below all other food

Q.36. What is the ideal temperature for pathogens to flourish?
- a) 10°C
- b) 37°C
- c) 55°C
- d) 90°C

Q.37. Which of the following is true about bacteria?
- a) Bacteria multiplies and grows faster in warm environments.
- b) Bacteria needs air to survive.
- c) Every type of bacteria can give people food poisoning.
- d) By freezing food, you can kill bacteria.

Q.38. How can you tell if food has enough bacteria to cause food poisoning?
- a) It will smell.
- b) You can't; it will appear normal.
- c) It will have a different colour.
- d) It will taste different.

Q.39. Which of the following do bacteria need to assist it to grow and multiply?
- a) Water
- b) Food
- c) Warm temperatures
- d) All of the above

Q.40. Which compounds are also mentioned as anti-nutritional factors (ANFs) (indicate all that apply)?
- a) Glycosides
- b) Trypsin inhibitors
- c) Saponins
- d) Glucosinolates
- e) Hemagglutinins

Q.41. What is the difference between permissible level and maximum residue level (MRL)?
- a) Permissible level gives the maximal contents of pesticides allowed in foods, while MRL gives the maximal acceptable concentration.
- b) There is no difference between permissible level and MRL.
- c) The permissible level is a temporary guideline, while the MRL is a permanent legal standard.
- d) The permissible level is the maximal acceptable concentration of pesticides in food, while MRL is the residue tolerance.

Q.42. When birds die of botulism the involved toxin inhibits release of ________.
- a) acetylcholine
- b) nicotine
- c) pepsin
- d) carboxylase
- e) sulfhydryl group

Q.43. What are the general toxic actions of lead resulting in anaemia (indicate all that apply)?
- a) Enhancement of Na-permeability
- b) Selective inhibition of resorption of nutrients
- c) Disturbance of heme synthesis
- d) Protoporphyria
- e) Agglutination of red blood cells

Q.44. Which of the following statements are true about the BMD 10 (indicate all that apply)?
a) The BMD10 takes the whole dose-response curve into account.
b) Fewer test animals result in a higher BMD10.
c) The BMD10 is not affected by the experimental design.
d) The BMD10 can be used as a starting point for calculating an ADI.
e) The BMD10 can be used to set a VSD by dividing it by a factor of 100.

Q.45. Zearalenone is a ________.
a) steroidal estrogenic
b) phytoestrogen
c) steroidal antiestrogen
d) non-steroidal estrogenic

Q.46. What foodborne exposure poses the greatest human-health risk worldwide?
a) Chemical contaminants/adulterants
b) Bacterial contamination
c) Mycotoxins and molds
d) Food additives

Answers (Exercise 1)

1. c	2. d	3. d	4. d	5. d	6. a	7. c & d	8. b	9. b	10. c
11. d	12. b	13. e	14. a	15. a & b	16. d	17. a	18. d	19. d	20. c
21. d	22. a	23. b	24. e	25. c	26. d	27. d	28. a	29. e	30. d
31. e	32. b	33. d	34. a	35. c	36. b	37. a	38. b	39. d	40. b & e
41. d	42. a	43. c & d	44. b, c & d	45. a	46. b				

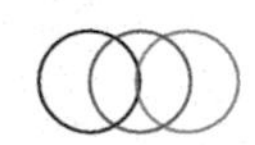

14.2 TRUE OR FALSE STATEMENTS

(Write T for True or F for False.)

Exercise 2

Q.1. Radiation can be introduced into the food chain naturally from cosmic rays entering the atmosphere.
Q.2. Nitrates and nitrites prevent the growth of *Salmonella*.
Q.3. Lectins and saponins are two plant sources of food poisoning.
Q.4. *E. coli* stains gram-negative because it has 90% peptidoglycan in its cell wall.
Q.5. *Staphylococcus aureus* is difficult to destroy because it is heat stable.
Q.6. HACCP is used to assess hazards by following the flow of foods.
Q.7. Virtually any food can serve as a vehicle for bacterial infection if handled carelessly by food workers.
Q.8. The vaccine that can be used for immunization against tetanus is *Claviceps purpura*.
Q.9. Spinal stimulation in botulism is caused mainly due to inhibition of glycine neurotransmitter.
Q.10. Botulin exotoxins are produced by *Claviceps purpurea* microorganism.
Q.11. Melamine is also used in some fertilizers.
Q.12. Cyanuric acid is a structural analogue of melamine.
Q.13. Melamine is used in the production of cyanuric acid, typically by reaction with formaldehyde.
Q.14. Cyanuric acid may be found as an impurity of melamine.
Q.15. Melamine and melanin are the same.
Q.16. Combination of melamine and cyanuric acid in diet is very safe.
Q.17. Ionophores are commonly added to ruminant feeds to increase milk production.
Q.18. Adult birds (chickens, turkeys, ratites) and birds with no previous exposure are resistant to ionophore coccidiostats.
Q.19. Food is more complex and variable in composition than all other substances to which humans are exposed.
Q.20. Nutrient substances include plant hormones and ingredients used in processed food.
Q.21. Non-nutrients are substances that may be classified as food additives.
Q.22. Water, chloride, fats (as micelles), short- and medium-chain fatty acids are transported through active transport.
Q.23. Two distinct types of colour additives have been approved for food use: those requiring certification by FDA chemists and those exempt from certification.
Q.24. Indirect food additives are substances that are *not* added directly to food but enter food by migrating from surfaces that contact food.
Q.25. Food hypersensitivity (allergy) is not an immune-mediated response.
Q.26. Foodborne illnesses do not involve any infection.
Q.27. ADI is generally based on results from animal toxicology studies.
Q.28. Supplements are regarded as foods or food constituents and not food additives or drugs.

Answers (Exercise 2)

1. T	2. F	3. T	4. F	5. T	6. T	7. T	8. F	9. T	10. F
11. T	12. T	13. F	14. T	15. F	16. F	17. F	18. F	19. T	20. F
21. T	22. F	23. T	24. T	25. F	26. F	27. T	28. T		

Exercise 3

Q.1. The toxic principle in avocado is ricin.
Q.2. Cows and goats are resistant to avocado toxicity.
Q.3. Chocolate is derived from the roasted seeds of the castor plant.
Q.4. Methylxanthines in chocolate varies because of the natural variation of cocoa beans and variation within brands of chocolate products.
Q.5. Unbaked yeast-containing bread dough is safe if a pet ingests it.
Q.6. Theobromine (3,7-dimethylxanthine) is safer than caffeine.
Q.7. Diazepam is contraindicated in poisoning by chocolate.
Q.8. Xylitol is also known as xylene.
Q.9. Theobromine is less toxic to rats, mice, and humans.
Q.10. HACCP is used to assess hazards by following the flow of foods.

Answers (Exercise 3)

1. F 2. F 3. F 4. T 5. F 6. F 7. F 8. F 9. T 10. T

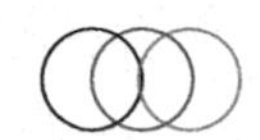

14.3 FILL IN THE BLANKS

Exercise 4

Q.1. Non-nutritive substances added to food to improve the physical, organoleptic, and nutritive properties or shelf life are called ________.
Q.2. Standards relating to food production and safety are covered by ________.
Q.3. Agents that prevent oxidative damage of food are called ________.
Q.4. The most toxic and suspected carcinogens among the food additives are ________.
Q.5. The rare adverse effect of the colouring agent tartrazine in humans is ________.
Q.6. The flavouring agent that produces the characteristic 'candy-shop' aroma and which is banned by the FDA due to its carcinogenic potential is ________.
Q.7. The flavour enhancing agent that produces the 'umami'-type tastes in foods is ________.
Q.8. The 'Chinese restaurant syndrome' is caused by the food additive ________.
Q.9. The oldest artificial sweetener used in foods is ________.
Q.10. The most common artificial sweetener approved by the FDA for use in pharmaceutical products and foods is ________.
Q.11. Vitamin C can react with benzoic acid, added as a preservative in foods, to form ________.
Q.12. The substances that are added to animal feeds to improve the quality of feeds or improve animal performance are called ________.
Q.13. Feed additives that increase the growth rate and feed conversion in animals are called ________.
Q.14. The most commonly used group of antibiotics approved for use as growth promoters in food-producing animals are ________.
Q.15. The most sensitive species of animal for ionophore toxicity is ________.
Q.16. The most common symptom in chronic ionophore antibiotic toxicity is ________.
Q.17. The approved antibiotic that should not be used simultaneously with other feed additives or growth promoters is ________.
Q.18. The use of zinc bacitracin is contraindicated in ________ animals.
Q.19. Apart from dietary sources, animals get exposed to urea from ________.
Q.20. Excessive use of ammonium nitrate and urea fertilizers in plants can result in ________ poisoning in animals.

Answers (Exercise 4)

1. Food additives
2. Codex Alimentarius Committee
3. Antioxidants
4. Colouring agents
5. Allergy or anaphylaxis
6. Safrole
7. Monosodium glutamate (MSG)
8. Monosodium glutamate (MSG)
9. Saccharin
10. Aspartame
11. Benzene
12. Feed additives
13. Growth promoters
14. Ionophore antibiotics
15. Horse
16. Cardiomyopathy

17. Zinc bacitracin
18. Lactating
19. Fertilizers
20. Nitrate

Exercise 5

Q.1. ________ is used in the production of melamine resins.
Q.2. Melamine is combined with ________ and other agents to produce melamine resins.
Q.3. Both melamine and cyanuric acid were relatively ________ when given individually.
Q.4. Horses are ________ to the toxic effects of ionophores in feed than cattle.
Q.5. Adult birds (chickens, turkeys, ratites) and birds with no previous exposure are more ________ to ionophore coccidiostats.
Q.6. After ingestion, non-protein nitrogen undergoes hydrolysis and releases excess ________ into the GI tract.
Q.7. Non-protein nitrogen (or NPN), which are not proteins, can be converted into proteins by ________ in the ruminant stomach.
Q.8. Apart from dietary sources, animals get exposed to urea from ________.
Q.9. Excessive use of ammonium nitrate and urea fertilizers in plants can result in ________. poisoning in animals.
Q.10. Ammonia exerts its toxic effect by inhibition of ________.

Answers (Exercise 5)

1. Melamine
2. Formaldehyde
3. Nontoxic
4. More sensitive
5. Sensitive
6. Ammonia (NH_3)
7. Microbes
8. Fertilizers
9. Nitrate
10. Citric acid cycle

Exercise 6

Q.1. The toxic principle in avocado is ________.
Q.2. ________ varieties of avocado have been most commonly associated with toxicosis.
Q.3. Animal studies show that exposure to persin leads to ________ in certain types of ________.
Q.4. Persin is an oil-soluble compound structurally similar to a ________.
Q.5. Chocolate is derived from the roasted seeds of ________.
Q.6. After high intake of avocado, goats and sheep develop severe ________ and ________.
Q.7. The most common victims of theobromine poisoning are ________.
Q.8. Domestic animals metabolize theobromine ________ than humans.
Q.9. Molds generally grow in stored fed stuffs containing a moisture content more than ________.
Q.10. The maximum permitted level of zearalenone in feed is ________.

 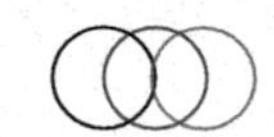

Answers (Exercise 6)

1. Persin
2. Guatemalan
3. Apoptosis, breast cancer cells
4. Fatty acid
5. *Theobroma cacao*
6. Mastitis, cardiac injury
7. Dogs
8. Slowly
9. 15%
10. 10 PPB

14.4 MATCH THE STATEMENTS

(Column A with Column B)

Exercise 7

Column A		Column B	
Q.1.	tetrodotoxin	a.	gluten (wheat)
Q.2.	ergotamine	b.	coral fish
Q.3.	ciguatoxin	c.	puffer fish
Q.4.	celiac disease/DON	d.	rye
Q.5.	organophosphate	e.	ADI
Q.6.	methyleugenol	f.	TDI
Q.7.	methylmercury	g.	MOE
Q.8.	E300 (ascorbic acid)	h.	ARfD

Answers (Exercise 7)

1. c   2. d   3. b   4. a   5. e   6. f   7. g   8. h

CHAPTER 15

MYCOTOXICOSES

15.1 MULTIPLE-CHOICE QUESTIONS

(Choose the most appropriate response.)

Exercise 1

Q.1. The maximum permitted level of zearalenone in feed is ________.
   a) 200 ppb
   b) 100 ppb
   c) 10 ppb
   d) 5 ppb

Q.2. The histological picture of uterus and vaginal zearalenone toxicosis is ________.
   a) hepatitis
   b) vaginitis
   c) nodule formation
   d) metaplasia

Q.3. Alimentary toxic aleukia (ATA) in human being is caused by ________ mycotoxins.
   a) DAS
   b) T2 toxin
   c) DON
   d) all of the above

Q.4. The mycotoxins that were used as biological warfare agents are ________.
   a) aflatoxins
   b) algal toxins
   c) trichothecenes
   d) fumonisins

Q.5. The species of animal that is more susceptible to trichothecenes is ________.
   a) tiger
   b) elephant
   c) horse
   d) cat

Q.6. An example for a neurotoxic mycotoxin is ________.
   a) aflatoxins
   b) trichothecenes
   c) tremorgens
   d) none of the above

Q.7. The most potent among the tremorgens is ________.
   a) ergotoxin
   b) penitrem A
   c) mycotoxin
   d) slaframine

Q.8. The most susceptible species for tremorgen mycotoxins are ________.
   a) cat and dog
   b) cattle and sheep
   c) poultry birds
   d) horse

Q.9. Ergotoxins are produced by the mold ________.
- a) *Claviceps purpurea*
- b) ryegrass
- c) trichothecenes
- d) none of the above

Q.10. Molds generally grow in stored food stuffs containing a moisture content more than ________.
- a) 25%
- b) 15%
- c) 50%
- d) 40%

Answers (Exercise 1)

1. c 2. d 3. d 4. c 5. d 6. c 7. b 8. a 9. a 10. b

 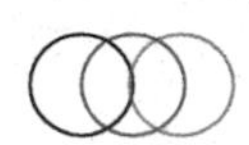

15.2 TRUE OR FALSE STATEMENTS

(Write T for True or F for False.)

Exercise 2

Q.1. A ruminant's facial eczema is caused by ochratoxins mycotoxicosis.
Q.2. The most potent nephrotoxic mycotoxins that cause mold nephrosis or mycotoxic nephropathy are sporidesmin.
Q.3. Ochratoxins are produced by *Aspergillus ochraceus* and *Penicillium viridicatum.*
Q.4. The most toxic among the ochratoxins is Ochratoxin A.
Q.5. The most susceptible species for ochratoxicosis are the horse and goat.
Q.6. The level of ochratoxin in feed should not exceed 50 ppb.
Q.7. An example for estrogenic mycotoxin, which can cause reproductive disorders, is zearalenone (f-2).
Q.8. Zearalenone (f-2) toxins are produced by the *Fusarium roseum* mold.
Q.9. The most susceptible species for zearalenone toxicosis is the elephant.
Q.10. Vulvovaginitis of hyperestrogenic syndrome in pigs is caused by the zearalenone mycotoxin.

Answers (Exercise 2)

1. F 2. F 3. T 4. T 5. F 6. F 7. T 8. T 9. F 10. T

15.3 FILL IN THE BLANKS

Exercise 3

Q.1. The secondary metabolites of fungus that cause deleterious effects to animals and human life is called ________.

Q.2. Molds generally grow in stored food stuffs containing a moisture content of more than ________.

Q.3. Mycotoxins are classified based on ________ affected.

Q.4. Turkey X disease is caused by mycotoxins ________.

Q.5. Aflatoxins are produced by ________ and ________.

Q.6. The most potent among the aflatoxins is ________.

Q.7. Aflatoxins are heat resistant but are unstable in ________.

Q.8. Aflatoxin-producing molds grow in stored feed stuffs that contain a moisture content more than ________.

Q.9. The domestic species of animal that is highly susceptible for aflatoxicosis are ________ and ________.

Q.10. The metabolite of aflatoxins that is excreted in milk and urine is ________.

Q.11. The aflatoxin content in cattle feeds should not exceed ________.

Q.12. The carcinogenic metabolite formed in the body from aflatoxins is ________.

Q.13. Aflatoxins causes defective protein synthesis by binding with ________ residue of DNA, causing a mispairing of nucleotides.

Q.14. Aflatoxin epoxide causes carcinogenic and mutagenic effect by causing ________ of the strands of DNA. (Alkylation forms cross-bridges between DNA strands.)

Q.15. Hemorrhage in aflatoxicosis is due to the decrease in ________ and ________.

Q.16. The type of carcinoma caused by aflatoxins is ________.

Q.17. Aflatoxins can be detected by ________ method.

Q.18. Hemorrhagic syndrome in poultry is caused by ________ mycotoxins.

Q.19. Rubratoxins are produced by ________ and ________.

Q.20. The most toxic metabolite of rubratoxin is ________.

Answers (Exercise 3)

1. Mycotoxins
2. 15%
3. Organ system
4. Aflatoxin
5. *Aspergillus flavus*, *A. parasiticus*
6. AFB1
7. UV light
8. 15%
9. Dog, ducklings
10. AFM1
11. 20 PPB
12. Aflatoxin 8, 9-epoxide
13. N-7 guanine
14. Alkylation
15. Prothrombin, vitamin K
16. Hepatocellular carcinoma
17. Thin layer chromatography

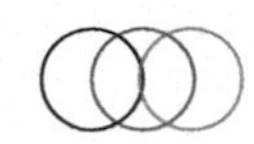

18. Rubratoxins
19. *Penicillium rubrum, P. purpurogenum*
20. Rubratoxin B

Exercise 4

Q.1. Feeding of ________ grass most commonly causes ergotism in cattle.
Q.2. Ergot alkaloids are partial agonists for ________ receptors.
Q.3. The ergot alkaloid that has an oxytocic effect on the uterus is ________ or ________.
Q.4. Acute egotism is manifested as a NERVOUS form while chronic ergotism is manifested as a ________ form.
Q.5. The type of ergotism commonly found in cattle is ________ form.
Q.6. In buffaloes, fusarium mycotoxins produce symptoms resembling chronic ergot poisoning known as ________.
Q.7. Botulin exotoxins are produced by ________ microorganism.
Q.8. Botulism is most common in ________ species.
Q.9. The foodborne toxicity caused due to preformed toxins is ________.
Q.10. Limber neck in birds due to paralysis of the neck muscles is caused by ________ toxins.
Q.11. The practice of using chicken manure as cattle feed or fertilizer can cause ________ toxicosis.
Q.12. Botulinum acts as a neurotoxin by inhibiting the release of ________ neurotransmitter.
Q.13. The specific treatment for botulism is ________.
Q.14. Tetanus toxin (also called tetanospasmin) is produced by ________.
Q.15. The most susceptible species of animal for tetanus is ________.
Q.16. 'Saw horse' or 'wooden horse' condition in horses is caused by ________.
Q.17. Spinal stimulation in botulism is caused mainly due to inhibition ________ neurotransmitter.
Q.18. The specific treatment for tetanus is administration of ________.
Q.19. For a pesticide based on toxicity studies, a NOAEL of 2 mg/kg bw/day in mice is derived. Calculate the maximal acceptable concentration in potatoes (if 200 g potatoes are eaten per day for a person weighing 60 kg) ________ mg/kg

Answers (Exercise 4)

1. Rye
2. Alpha
3. Ergometrine or ergonovine
4. Gangrenous
5. Gangrenous ergotism
6. Degnala disease
7. *Clostridium botulinum*
8. Chicken
9. Botulism
10. Botulinum
11. Botulinum
12. Acetylcholine
13. Polyvalent botulinum antitoxin
14. *Clostridium tetani*
15. Horse
16. Tetanus
17. Glycine
18. Tetanus antitoxin
19. 6 mg/kg product

Exercise 5

Q.1. A ruminant's facial eczema is caused by ________ mycotoxicosis.
Q.2. The most potent nephrotoxic mycotoxins that cause mold nephrosis or mycotoxic nephropathy are ________.
Q.3. Ochratoxins are produced by ________ and ________.
Q.4. The most toxic among the ochratoxins is ________.
Q.5. The most susceptible species for ochratoxicosis are ________ and ________.
Q.6. The level of ochratoxin in feed should not exceed ________.
Q.7. An example for estrogenic mycotoxin that can cause reproductive disorders is ________.
Q.8. Zearalenone (F-2) toxins are produced by ________ mold.
Q.9. The most susceptible species for zearalenone toxicosis is ________.
Q.10. Vulvovaginitis of hyperestrogenic syndrome in pigs is caused due to ________ mycotoxin.
Q.11. The maximum permitted level of zearalenone in feed is ________.
Q.12. The histological picture of uterus and vaginal zearalenone toxicosis is ________.
Q.13. Alimentary toxic aleukia (ATA) in human beings is caused by ________ mycotoxins.
Q.14. The mycotoxins that were used as biological warfare agents are ________.
Q.15. The species of animal that is more susceptible to trichothecenes is ________.
Q.16. An example for neurotoxic mycotoxin is ________.
Q.17. The most potent among the tremorgens is ________.
Q.18. The most susceptible species for tremorgen mycotoxins are ________ and ________.
Q.19. Staggers syndrome in cattle is produced by ________ mycotoxins.
Q.20. Ergotoxins are produced by the mold ________.

Answers (Exercise 5)

1. Sporidesmin
2. Ochratoxins
3. *Aspergillus ochraceus* and *Penicillium viridicatum*
4. Ochratoxin A
5. Birds and pig
6. 10 PPB
7. Zearalenone (F-2)
8. *Fusarium roseum*
9. Pig
10. Zearalenone
11. 10 PPB
12. Metaplasia
13. Trichothecenes
14. Trichothecenes
15. Cat
16. Tremorgens
17. Penitrem A
18. Cattle and dog
19. Tremorgen
20. *Claviceps purpurea*

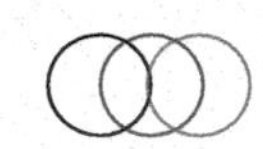

15.4 MATCH THE STATEMENTS

(Column A with Column B)

Exercise 6

	Column A		Column B
Q.1.	ergotism in cattle	a.	tetanus
Q.2.	ergot alkaloids are partial agonist	b.	alpha receptors
Q.3.	tetanus toxin	c.	horse
Q.4.	'saw horse' or 'wooden horse'	d.	rye (*C. purpurea*)
Q.5.	form of ergotism commonly found in cattle	e.	gangrenous
Q.6.	most susceptible species of animal for tetanus	f.	*Clostridium tetani*
Q.7.	tetanus	g.	chicken
Q.8.	botulinum action	h.	acetylcholine
Q.9.	oxytocic effect on uterus	i.	ergometrine or ergonovine
Q.10.	botulism is most common	j.	tetanus antitoxin

Answers (Exercise 6)

1. d 2. b 3. f 4. a 5. e 6. c 7. j 8. h 9. i 10. g

Section 7
ENVIRONMENTAL TOXICOLOGY

Chapter 16

POLLUTION AND ECOTOXICOLOGY

16.1 MULTIPLE-CHOICE QUESTIONS

(Choose the most appropriate response.)

Exercise 1

Q.1. A threshold-limit value-time-weighted average for a chemical represents _______.
- a) an airborne concentration of a chemical that can never be exceeded
- b) an airborne concentration of a chemical that is believed to cause no adverse health effect to a worker exposed for 8 hours a day, 40 hours a week
- c) an airborne concentration of a chemical that cannot be exceeded for longer than 15 minutes a day
- d) a value for an acceptable airborne concentration of a chemical established by the Occupational Safety and Health Administration
- e) an airborne concentration which cannot be measured using available technology

Q.2. Which of the following is the most significant contributor to air pollution by mass in sub-urban areas?
- a) Manufacturing
- b) Transportation
- c) Space heaters
- d) Electric power generation
- e) Waste disposal

Q.3. The major toxic effect of hydrogen cyanide exposure is _______.
- a) lung damage
- b) haemoglobin alteration
- c) haemolysis of RBCs
- d) inhibition of mitochondrial respiration
- e) lipid peroxidation

Q.4. Which of the following agents would *not* likely produce reactive airways dysfunction syn-drome (RADS)?
- a) Carbon monoxide
- b) Chlorine
- c) Ammonia
- d) Toluene di-isocyanate
- e) Acetic acid

Q.5. Benzene is similar to toluene _______.
- a) in its metabolism to redox active metabolites
- b) regarding covalent binding of its metabolites to proteins
- c) in its ability to produce CNS depression
- d) in its ability to produce acute myelogenous leukaemia
- e) in its ability to be metabolized to benzoquinone

Q.6. Which of the following is *not* a primary pollutant?
- a) Carbon monoxide
- b) Lead
- c) Ozone
- d) Nitrogen dioxide

Q.7. Most automobiles emit up to _______ % less pollutants now than in 1960.
a) 80
b) 30
c) 99
d) 50

Q.8. Particulates may cause which of the following health hazards to humans?
a) Respiratory distress
b) Damage to nervous system
c) Blindness
d) Learning disabilities

Q.9. Carbon monoxide may produce one of the symptoms of exposure in humans:
a) Excess urination
b) Headache
c) Throat irritation
d) Hearing loss

Q.10. Tropospheric ozone may cause which of the following effects?
a) Grime deposits
b) Reduced visibility
c) Retardation of plant growth
d) Metal corrosion

Q.11. Fine particulates from motor vehicles and power plants are reported to kill about _______ Americans annually.
a) 38,000
b) 2 million
c) 9100
d) 64,000

Q.12. _______ may be the primary health problem when a person's bronchial tubes respond to allergens, pollution, etc. resulting in hyperactive airways.
a) Cardiac arrest
b) Asthma
c) Laryngitis
d) Sneezing

Q.13. Signs of an extended asthma attack may include _______.
a) sweating
b) rapid pulse
c) skin turns blue
d) all of the above

Q.14. Most adults spend an average of _______ % of their time indoors.
a) 50
b) 25
c) 90
d) 10

Q.15. Which is a potential source of indoor air pollution?
a) Moisture
b) Room air fresheners
c) Personal care products
d) All of the above

Q.16. A smoker is exposed to nearly _______ compounds in mainstream cigarette smoke.
a) 4700
b) 320
c) 9400
d) 560

 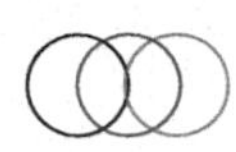

Q.17. Building related illnesses (BRI) refers to _______.
- a) well-defined illnesses occurring in a building that can be traced to specific building problems
- b) the display of acute symptoms by a number of people in a building without a particular pattern and the varied symptoms cannot be associated with a particular pattern
- c) well-defined illnesses occurring in a building that cannot be traced to specific building problems
- d) the display of acute symptoms by a number of people in a building with a specific pattern of disease associated with a particular pattern

Q.18. Biological contaminants are most likely aggravated by what problem?
- a) Auto exhaust
- b) Unvented gas stove
- c) Moisture
- d) Household chemicals

Q.19. Air that is drawn into the home by cracks in the foundation is known as _______.
- a) natural ventilation
- b) infiltration
- c) mechanical filtration
- d) foundation suction

Q.20. When considering the contribution of all greenhouse gases, carbon monoxide contributes approximately _______ % to global warming?
- a) 24
- b) 6
- c) 55
- d) 15

Q.21. What event in 1978 was (one of) the first that raised public consciousness focusing attention on the hazards to the environment and human health of improperly disposed chemicals?
- a) Love canal
- b) Three-mile island
- c) Chernobyl
- d) Valdez oil spill
- e) none of them

Q.22. _______ is the leading factor among those listed below, associated with cancer risk.
- a) occupation
- b) alcohol
- c) pollution
- d) medicines and medical procedures
- e) poor diet

Q.23. Which of the following diseases is an acute disease?
- a) AIDs
- b) Emphysema
- c) Cancer
- d) Flu
- e) Heart disease

Q.24. The leading cause of injury-related death in the United States is _______.
- a) shootings
- b) falls
- c) motor vehicles
- d) stabbings
- e) other forms of trauma to the human body

Q.25. What is the major mechanism for chemical injury caused by allergens?
a) Interference with enzyme activity
b) Directly combing with some cell component other than enzymes
c) Producing a secondary action in which a chemical causes the release or formation of a more harmful substance
d) None of the above
e) All of the above

Q.26. Which of the following chemicals is most associated with systemic disease involving the central nervous system, the gastrointestinal system, and the blood-forming tissues?
a) Lead
b) Pollen
c) Animal dander
d) CO
e) None of the above
f) All of the above

Q.27. Which of the following is *not* an exogenous factor of malignant tumours?
a) Habits
b) Ionizing radiation
c) Chemical exposure
d) Environment (socioeconomic, geographical, and occupational)
e) Gender

Q.28. Which of the following is *not* an endogenous factor of malignant tumours?
a) Oncogenic viruses
b) Gender
c) Age
d) Hormonal imbalance
e) Impaired immune system

Q.29. How many homologous pairs of chromosomes are there in the human nucleus?
a) 20
b) 21
c) 22
d) 23
e) 24

Q.30. Which of the following diseases is due to the point mutation on the co-dominant genes resulting in abnormal haemoglobin?
a) Phenylketonuria (PKU)
b) Cystic fibrosis
c) Sickle cell disease
d) Huntington's disease
e) Spherocytosis

Answers (Exercise 1)

1. b	2. b	3. d	4. a	5. c	6. c	7. a	8. a	9. b	10. c
11. d	12. b	13. d	14. c	15. d	16. a	17. a	18. c	19. b	20. c
21. a	22. e	23. d	24. c	25. c	26. a	27. e	28. a	29. d	30. c

Exercise 2

Q.1. How many synthetic chemicals are currently in commercial use in the United States, and whose toxicity is *not* widely known or understood?
   a) 70,000
   b) 6000
   c) 500
   d) 800,000
   e) 15,000

Q.2. _______ is the most likely process of absorption for amino acids.
   a) Diffusion
   b) Facilitated diffusion
   c) Active transport
   d) Endocytosis
   e) None of the above

Q.3. Which of the following sites in the respiratory system is the most likely place for the carbon dioxide and oxygen to exchange in the blood?
   a) Nose
   b) Pharynx
   c) Larynx
   d) Trachea
   e) Alveoli

Q.4. Which of the following processes, when prolonged and severe, can be life threatening such as in asthmatic attacks?
   a) Mucociliary streaming
   b) Coughing
   c) Sneezing
   d) Bronchoconstriction
   e) None of the above

Q.5. What is the best estimate for the area of skin coverage in the average adult?
   a) 2500 in^2
   b) 3000 in^2
   c) 3500 in^2
   d) 4000 in^2
   e) 4500 in^2

Q.6. By what absorptive process does hexane pass through the skin?
   a) Passive diffusion
   b) Facilitated diffusion
   c) Active transport
   d) Endocytosis
   e) None of the above

Q.7. How long is the average adult human gastrointestinal tract?
   a) 20 feet
   b) 22 feet
   c) 30 feet
   d) 41 feet
   e) 52 feet

Q.8. Where is the most likely site for the absorption of toxic agents in the gastrointestinal tract?
   a) Between the stomach and the upper portion of the intestine
   b) Stomach
   c) Small intestine
   d) Large intestine
   e) The lower portion of large intestine

Q.9. What is the mechanism for the harmful effects of carbon monoxide (CO)?
  a) Interfere with or block the active sites of some important enzymes
  b) Direct chemical combination with a cell constituent
  c) Secondary action as a result of its presence in the system
  d) Compete with the co-factors for a site on an important enzyme
  e) None of the above

Q.10. What is the major organ responsible for detoxification in the body?
  a) Lung
  b) Intestines
  c) Kidney
  d) Liver
  e) Skin

Q.11. Which of the following might be linked to parkinsonism?
  a) Nitrogen dioxide
  b) Zinc
  c) Copper
  d) Magnesium
  e) Carbon monoxide

Q.12. Which of the following compounds is *not* an oxidant-type air pollutant?
  a) NO_2
  b) SO_2
  c) O_3
  d) Radical hydrocarbons
  e) Aldehydes

Q.13. Which of the following pollutants contributes most to non-tobacco-smoking lung cancer?
  a) Asbestos
  b) Vinyl chloride
  c) Benzene
  d) Products of incomplete combustion
  e) Formaldehyde

Q.14. Inhalants, such as NO_2 and trichloroethylene, can increase proliferation of opportunistic pathogens in the lungs by _______.
  a) destroying goblet cells in the respiratory tract
  b) damaging the alveolar septa
  c) inactivating cilia in the respiratory tract
  d) killing alveolar macrophages
  e) dampening the immune system

Q.15. Which of the following is *not* a characteristic of SO_2 toxicology?
  a) SO_2 is a major reducing-type air pollutant.
  b) Increased air flow rate increases the amount of SO_2 inhaled.
  c) SO_2 inhalation causes vasoconstriction and increased blood pressure.
  d) SO_2 is predominately absorbed in the conducting airways.
  e) SO_2 inhalation increases mucus secretion in humans.

Q.16. Which of the following would be *most* likely to occur on sulfuric acid exposure?
  a) Vasoconstriction
  b) Decreased mucus secretion
  c) An anti-inflammatory response
  d) Vasodilation
  e) Bronchoconstriction

 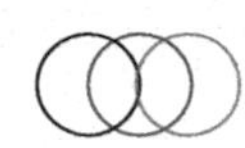

Q.17. All of the following statements regarding particulate matter are true *except* _______.
a) Metals are most commonly released into the environment during coal and oil combustion.
b) The interaction of gases and particles in the atmosphere can create a more toxic product than the gas or particles alone.
c) Solubility does not play a role in the bioavailability of a metal.
d) The earth's crust is an important source of atmospheric magnesium.
e) Diesel exhaust contains reducing- and oxidant-type air pollutants.

Q.18. Which of the following statements is *not* true?
a) Ozone (O_3) combines with a nitric-oxide radical to form NO_2.
b) O_2 combines with an oxygen radical to form ozone.
c) O_3 can cause damage to the respiratory tract.
d) Accumulation of O_3 in the stratosphere is important for protection against UV radiation.
e) Cl_2 gas is known to cause O_2 degradation.

Q.19. Which of the following is *not* a likely symptom of NO_2 exposure?
a) Increased secretion by Clara cells
b) Pulmonary edema
c) Shortness of breath
d) Loss of ciliated cells in bronchioles
e) Decreased immune response

Q.20. Which of the following statements regarding aldehyde exposure is *false*?
a) The major aldehyde pollutants are formaldehyde and acrolein.
b) Formaldehyde is found in tobacco smoke, but acrolein is not.
c) Acrolein causes increased pulmonary flow resistance.
d) Formaldehyde exposure induces bronchoconstriction.
e) The water solubility of formaldehyde increases its nasopharyngeal absorption.

Q.21. Carbon monoxide (CO) exerts its toxic effects via its interaction with which of the following?
a) DNA polymerase
b) Actin
c) Kinesin
d) Haemoglobin
e) Microtubules

Q.22. What is the mode by which a chemical enters the lithosphere?
a) Evaporation
b) Adsorption
c) Dissolution
d) Absorption
e) Diffusion

Q.23. The bioavailability of contaminants in the hydrosphere is directly related to _______.
a) chemical concentration
b) amount of chemical
c) water solubility of chemical
d) toxicity of chemical
e) molecular size of chemical

Q.24. All of the following regarding biomarkers are true *except* _______.
a) Dermal absorption is considered an external dose.
b) Biomarkers of susceptibility are useful in extrapolating wildlife disease to human diseases.
c) Induction of certain enzymes is an important biomarker.

d) The biologically effective dose is the amount of internal dose needed to elicit a certain response.
e) The effects of chemical exposure can be different across species.

Q.25. Which of the following processes is *least* likely to be affected by endocrine-disrupting agents?
a) Enzyme activity
b) Transcription
c) Hormone secretion
d) Signal transduction
e) DNA replication

Q.26. Estrogen exposure has been shown to cause all of the following in wildlife species *except* _______.
a) sexual imprinting
b) altered sex hormone levels
c) immune suppression
d) gonadal malformations
e) sex reversal

Q.27. Which of the following is *false* regarding terrestrial ecotoxicology?
a) Terrestrial organisms are generally exposed to contaminants via ingestion.
b) Predation is an important confounder of measurements in terrestrial toxicology field studies.
c) Reproductive tests are not important in measuring end points in toxicity tests.
d) Enclosure studies are better able to control for environmental factors in field studies.
e) Toxicity tests usually test the effects of an oral chemical dose.

Q.28. An important type(s) of compound that is far more toxic in water than in air is/are _______.
a) organic compounds
b) photochemicals
c) vapours
d) lipid-soluble xenobiotics
e) metals

Q.29. Which of the following are used to record end-point toxicity of aquatic toxicity tests?
a) LD_{50} and ED_{50}
b) LC_{50} and EC_{50}
c) Reproductive tests
d) LD_{50} and LC_{50}
e) LD_{50} and EC_{50}

Q.30. Biologic availability is _______.
a) the total amount of chemical within an organism
b) the concentration of chemical in an environmental reservoir
c) the threshold concentration of a chemical needed for toxic effect
d) the concentration of chemical within an organism
e) the proportion of chemical potentially available for uptake

Q.31. Chemodynamics does *not* study _______.
a) the fate of chemicals in the environment
b) the rate at which chemicals are metabolized
c) the distribution of chemicals in the environment
d) the effects of toxic substances on the environment
e) the release of chemicals into the environment

Answers (Exercise 2)

1. a	2. b	3. e	4. d	5. b	6. a	7. c	8. a	9. b	10. d
11. e	12. b	13. d	14. e	15. c	16. e	17. c	18. d	19. a	20. b
21. d	22. b	23. c	23. a	25. e	26. c	27. c	28. e	29. b	30. e
31. d									

Exercise 3

Q.1. What is the major toxic mechanism for hydrogen cyanide?
- a) Interfere with or block the active sites of the enzyme
- b) Inactivate or remove the co-factor
- c) Compete with the co-factor for a site on the enzyme
- d) Altering enzyme structure directly thereby changing the specific three-dimensional nature of the active site
- e) None of the above

Q.2. What is the major mechanism for toxicity of dithiocarbamate during alcohol consumption?
- a) Interfere with or block the active sites of the enzyme
- b) Inactivate or remove the co-factor
- c) Compete with the co-factor for a site on the enzyme
- d) Altering enzyme structure directly thereby changing the specific three-dimensional nature of the active site
- e) None of the above

Q.3. Which of the following substances can cause a syndrome in infants referred to as 'blue baby'?
- a) Carbon monoxide
- b) Chlorine gas
- c) Ozone
- d) Sulfuric oxides
- e) Nitrogen compounds

Q.4. Which of the following refers to a substance that is attached to an antigen and promotes an antigenic response?
- a) Light chain
- b) Heavy chain
- c) Leukocyte
- d) Helper cell
- e) Hapten

Q.5. Which kind of cells are the primary targets of the AIDS virus?
- a) Cytoxic (killer) cell
- b) Helper T-cells (e.g., CD4)
- c) Memory cells
- d) Suppressor T-cells
- e) Delayed hypersensitivity T-cells

Q.6. The largest percent of antibodies belong to the _______ class.
- a) IgG
- b) IgE
- c) IgM
- d) IgA
- e) IgD

Q.7. Which of the following hypersensitivity reactions is most often seen in transfusion reactions?
- a) Cytotoxic
- b) Cell-mediated

c) Immune complex
d) Anaphylactic
e) None of the above

Q.8. What kind of hypersensitivity is associated with asthma?
a) Cytotoxic
b) Cell-mediated
c) Immune complex
d) Anaphylactic
e) None of the above

Q.9. What kind of the following interactions is characteristic of that for caffeine and sleeping pills?
a) Additive
b) Synergistic
c) Antagonistic
d) None of the above
e) They don't interact with each other

Q.10. Yu-Cheng ('oil disease') in Taiwan is due to the toxic effect of _______.
a) lead
b) PCBs
c) dioxin
d) asbestos
e) mercury

Q.11. When were both the federal regulatory and legislative efforts begun to reduce lead hazards, including the limitation of lead in paint and gasoline?
a) 1970s
b) 1980s
c) 1990s
d) 1940s
e) 1950s

Q.12. Under the CAAA of 1990, an allowance is the right to emit how much sulfur dioxide?
a) 100 tons
b) 10 tons
c) 1000 tons
d) 1 ton

Q.13. Which of the following is *not* an indicator pollutant for regulation under the National Ambient Air Quality Standards (NAAS) provisions?
a) Sulfur dioxide (SO_2)
b) Carbon monoxide (CO)
c) Nitrogen oxides (NOx)
d) Asbestos
e) Particulate matter (PM-10)

Q.14. Reformulated gasoline ('oxygenated fuel') with a 2 percent minimum oxygen content is required during the winter months in non-attainment areas for carbon monoxide. Which of the substances listed below may be added to fuel to render it oxygenated?
a) Lead
b) Methyl tertiary butyl ether
c) Organic magnesium
d) Mercury
e) Benzene

Q.15. The federal Clean Water Act (CWA) and its amendments apply primarily to _______.
a) protection of surface waters
b) protection of groundwater

 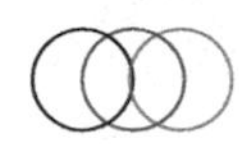

   c) navigable waterways
   d) boat moorings
   e) bays and estuaries

Q.16. Runoff that results from rain falling on roofs, roads, parking lots, loading docks, storage areas, and other areas exposed to rain is referred to as _______.
   a) storm water
   b) effluent
   c) dirty water
   d) fugitive effluent
   e) escaped liquid

Q.17. 'Source reduction' refers to the deliberate decrease in the amounts of any hazardous substance, contaminant, or pollutant that enters the environment prior to recycling, treatment, or disposal. This practice is most closely associated with what federal regulation?
   a) Resource Conservation and Recovery Act (RCRA)
   b) Clean Water Act
   c) Clean Air Act Amendments of 1990
   d) Comprehensive Environmental Response, Compensation and Liability Act (CERCLA)
   e) Pollution Prevention Act of 1990

Q.18. Many companies are encouraged by the USEPA and State Environmental Protection Agencies to follow certain policies in dealing with environmental pollution. This now normally involves the use of _______.
   a) really good lawyers
   b) concealment
   c) cleanup
   d) pollution prevention
   e) witness protection programs

Q.19. There are several elements to a proactive environmental management program. Which of one of the following does *not* apply?
   a) Hold orientation and training sessions where environmental policy can be communicated to every employee.
   b) Establish clear lines of authority with written policies for compliance and corrective measures with prompt reporting.
   c) Reporting requirements and schedules for self-reporting data to regulatory agencies should be monitored.
   d) Avoid providing environmental information to employees and management so they can claim they were unaware of the law.
   e) Develop a computerized information management system to coordinate corporate-wide data to identify problems, evaluate compliance, and target opportunities for future compliance planning.

Q.20. Permissible Exposure Limits are _______.
   a) promulgated by the Occupational Safety and Health Administration and have the force of law
   b) identical to and interchangeable with Short Term Exposure Limits
   c) authorized under the Toxic Substances Control Act
   d) promulgated by the Environmental Protection Agency
   e) are directly adopted from the American Conference of Governmental Industrial Hygienists Threshold Limit Value list

Q.21. What is the process by which plants use energy from the sun to turn carbon dioxide and water into simple sugars?
   a) Photosynthesis
   b) Breathing
   c) Retrogradation

d) Retrogression
e) Photographic

Q.22. What is the best example(s) from the list below of an organism belonging to the first trophic level?
a) Human
b) Wolf
c) Plant
d) Large cat
e) a and c

Q.23. Give an example of a primary consumer.
a) Caterpillar
b) Grasshopper
c) Cattle
d) Elephants
e) All of the above

Q.24. Circle three macronutrients listed below.
a) Sulfur
b) Copper
c) Carbon
d) Oxygen
e) Iron

Q.25. Circle all the ways carbon dioxide is released into the atmosphere.
a) Respiratory process of animals
b) Combustion of fossil or organic fuels
c) Decomposition of organic matter
d) The emissions of electric cars
e) The heavy use of phones

Q.26. What gas makes up the largest percent of the earth's air?
a) Iron
b) Nitrogen
c) Oxygen
d) Carbon
e) Phosphorus

Q.27. Which substance has been identified as a respiratory tract carcinogen in humans?
a) Kaolin
b) Hydrogen fluoride
c) Arsenic
d) Cotton dust
e) Vanadium

Q.28. Chloracne is associated with _______.
a) prominent hyperkeratosis of the follicular canal
b) production of excessive sebum
c) exposure to halogenated aliphatic hydrocarbons
d) exposure to chlorine gas
e) increases in serum androgen levels

Q.29. Which of the following statements is *not* true?
a) Arsenic, benzene, and vinyl chloride are known as human carcinogens.
b) Many peroxisome proliferators cause hepatic tumours in rats and are promoting agents for hepatocarcinogenesis.
c) Benzidine, beta-naphthylamine, and derived dyes have caused urinary bladder tumours in exposed workers.

 d) Short asbestos fibres (<2 μm long) are thought to be predominately responsible for the induction of mesotheliomas.
 e) Butylated hydroxyanisole (BHA) acts as a non-mutagenic carcinogen in the forestomach of rats.

Q.30. Which of the following is characteristic of a non-genotoxic carcinogen?
 a) Has no influence on the promotional stage of carcinogenesis
 b) Would be expected to produce positive responses in in-vitro assays for mutagenic potential
 c) Typically exerts other forms of toxicity and/or disrupts cellular homeostasis
 d) Generally shows little structural diversity
 e) Typically has little effect on cell turnover

Q.31. The Resource Conservation and Recovery Act (RCRA) applies to facilities and agencies that do which of the following with hazardous waste?
 a) Generate and store
 b) Dispose of
 c) Treat
 d) Transport
 e) All of the above

Q.32. Under RCRA definition, a waste that explodes or reacts with water or acid and is unstable, is considered to be ________.
 a) ignitable
 b) corrosive
 c) reactive
 d) toxic
 e) none of the above

Q.33. Off-site shipments of hazardous waste must be labelled and marked according to requirements of what agency?
 a) Environmental Protection Agency (EPA)
 b) Department of Consumer Affairs
 c) Department of Defense (DOD)
 d) Department of Transportation (DOT)
 e) Department of Energy (DOE)

Q.34. A step in the Superfund process is to identify 'PRPs' that can be required to finance cleanup activities. PRPs refer to:
 a) Potentially Remaining Parties
 b) Possible Responsible Polluters
 c) Potentially Responsible Parties
 d) Potentially Remaining Polluters
 e) None of the above

Q.35. If you are responsible for, or aware of, a substance that has been discharged or released into the environment in an amount that exceeds its listed 'R' under SARA Title III, what must you do according to these regulations?
 a) Immediately report the release to local emergency response agencies and state and national emergency response agencies
 b) Seek out an attorney
 c) Deny knowledge of the incident
 d) Take cover
 e) Rinse it down with a garden hose

Q.36. UN-based packaging regulations divide hazardous materials into Groups I, II, or III. The most hazardous materials fall into which group ________?
 a) Group I
 b) Group II

c) Group III
d) Non-conforming group
e) None of the above

Q.37. According to a federal Underground Storage Tank (UST) law found under RCRA, tanks storing petroleum or hazardous chemicals are required to have leak-detection systems installed by what date _______?
a) 1995
b) 2000
c) Whenever convenient
d) 1993
e) Does not apply to USPS

Q.38. If only 10% of the volume of petroleum is contained in pipes underground attached to the tank, does the system still qualify as a UST under federal regulations?
a) Yes
b) No

Q.39. These actions may be covered by a portion of tort law that deals with acts not intended to inflict injury but where persons may be harmed by the careless and improper actions of another, such as the improper disposal of hazardous wastes.
a) Knowing endangerment
b) Negligent actions
c) Knowing actions
d) Willful avoidance

Q.40. Environmental law is a system of laws that encompass all of the environmental protections that originate from all the sources listed below *except* _______.
a) United States constitution and state constitutions
b) regulations published by federal, state, and local agencies
c) justice of the Peace rulings
d) presidential executive orders the common law

Q.41. What point source is the most significant contributor to air pollution by mass in suburban areas?
a) Cattle farms
b) Transportation
c) Electric power generation
d) Waste disposal

Q.42. What was the regulatory response to the Delaney clause?
a) Permitted most food additives to be declared generally recognized as safe (GRAS)
b) Prohibited EPA from setting safe exposure levels for environmental carcinogens
c) Prohibited FDA from approving food additives found to cause cancer in animals
d) Was applied only to food additives demonstrating human evidence of carcinogenicity

Q.43. What does a threshold limit value-time weighted average (TLV-TWA) for a chemical represent?
a) An airborne concentration of a chemical that can never be exceeded during an 8-hour workday
b) A mean airborne concentration of a chemical believed to cause no adverse health effects to workers exposed for 8 hours/day, 40 hours/week
c) An airborne concentration of a chemical that cannot be exceeded for longer than 15 minutes during an 8-hour workday
d) An acceptable mean airborne concentration of a chemical established by the occupational safety and health administration (OSHA)

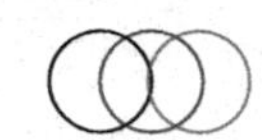

Answers (Exercise 3)

1. b	2. b	3. e	4. e	5. b	6. a	7. a	8. d	9. c	10. b
11. a	12. d	13. d	14. b	15. a	16. a	17. e	18. d	19. d	20. a
21. a	22. c	23. e	24. a, c & d	25. a, b & c	26. b	27. c	28. a	29. d	30. c
31. e	32. c	33. d	34. c	35. a	36. a	37. d	38. a	39. b	40. c
41. b	42. c	43. b							

16.2 TRUE OR FALSE STATEMENTS

(Write T for True or F for False.)

Exercise 4

Q.1. An inversion is a stable, slow-moving air mass that results from the formation of a cool layer of air above warmer air near the earth.
Q.2. Global warming can be attributed to greenhouse gases.
Q.3. Ozone depletion has been linked to chlorofluorocarbons.
Q.4. There are traces of helium found in the atmosphere.
Q.5. The combined reflective ability of cloud cover and ground surfaces are known as inversions.
Q.6. The troposphere is the layer of air in the 20–40-km altitude range.
Q.7. Ozone is a primary pollutant.
Q.8. The amount of nitrogen dioxide in the atmosphere has decreased in the last 20 years.
Q.9. Lead has been reduced by 98% in ambient concentrations from 1970.
Q.10. An oxidant is a substance that removes hydrogen from a compound.
Q.11. Carbon monoxide is a gray-coloured gas with an odour like rotten eggs.
Q.12. Nasal hairs play a part in cleaning air before it reaches the lungs.
Q.13. The beginning signs of an asthma attack include sweating and rapid pulse.
Q.14. During an airway spasm, the muscles of the alveoli contract, and the membranes swell.
Q.15. There is a phenomenon called sick building syndrome.
Q.16. Cat saliva is considered to be a biological contaminant.
Q.17. The number of asthma cases since 1982 have decreased.
Q.18. The New England area is considered to be an acid-sensitive region.
Q.19. The Kyoto conference in 1997 suggested that the United States decrease greenhouse-gas emissions 7% below 1990 levels.
Q.20. Acid deposition is a threat to aquatic ecosystems because it leads to reductions in fish populations.

Answers (Exercise 4)

1. F	2. T	3. T	4. T	5. F	6. F	7. F	8. F	9. T	10. T
11. F	12. T	13. F	14. F	15. T	16. T	17. F	18. T	19. T	20. T

Exercise 5

Q.1. Carcinogens refers to chemicals that primarily cause reproductive disorders.
Q.2. Most cancer risks are modifiable through changes in human behaviour.
Q.3. The best estimate for the contribution of pollution to cancer risk is around 2% of the total risk.
Q.4. Chemical pollutants contribute substantially to cancer risk.
Q.5. Reservoirs are those living organisms or inanimate objects that provide the conditions where the organisms may survive, multiply, and also provide the conditions necessary for transmission.
Q.6. The leading cause of injury-related death in the United States involve motor vehicles accounting for nearly 4,000 deaths per year with more than 40% of those deaths in the 16–19 years' age group.
Q.7. Falls are the leading causes of accidental deaths in the home.
Q.8. Teratologic defects usually arise during the embryonic period of development, and the causative factors are usually not genetic but from exposure to chemicals or radiation.
Q.9. Kwashiorkor develops from a lack of vitamin C.

 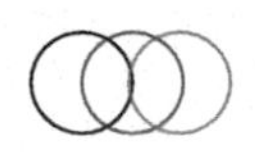

Q.10. Environmental disease refers to any pathologic process having a characteristic set of signs and symptoms that are detrimental to the well-being of the individual and are the consequence of many factors, including exposure to physical or chemical agents, poor nutrition, and social or cultural behaviours.
Q.11. Protein is the basic informational macromolecule that is the basis of heredity whose chemical structure has been known for more than 45 years.
Q.12. Since every individual within each species is very much a representation of the information contained in the genome together with the expression of those genes in the development of that individual, all the genotypic changes are harmful or damaging.
Q.13. One of the most well-known cytogenetic defects is trisomy 21 or Down's disease, which is characterized by the addition of an extra chromosome to chromosome number 21.
Q.14. Most teratogens exert their effect during certain critical time windows in the late stages of tissue and organ formation known as organogenesis.
Q.15. Lead has been recognized as a hazard since early civilization when it was used to store wine, to pipe water, and even as vessels in which to cook food.
Q.16. Once a potential toxic substance goes into our society, it automatically produces an adverse effect.
Q.17. External respiration refers to the exchange of gases between blood and individual cells.
Q.18. Sulfur oxides tend to reach deep into lung tissue while nitrogen dioxides tend to act in the upper moist airways of the respiratory tree.
Q.19. The skin is the body's largest organ and consists of many interconnected tissues.
Q.20. Epidermis is the outer, thinner layer of the skin, and the dermis is the inner and much thicker layer of the skin.

Answers (Exercise 5)

1. F	2. T	3. T	4. F	5. T	6. F	7. T	8. T	9. F	10. T
11. F	12. F	13. T	14. F	15. T	16. F	17. F	18. F	19. T	20. T

Exercise 6

Q.1. Zinc, iron, and copper are examples of macronutrients.
Q.2. Humans are nitrogen-based life forms.
Q.3. The sedimentary cycle refers to calcium, iron, and phosphorus getting leached from sedentary rocks by water erosion.
Q.4. The gastrointestinal tract is a major route of absorption for many toxic agents including mercury, lead, and cadmium.
Q.5. A toxin can produce a harmful effect upon an organ only by stimulating the normal metabolic actions of that particular organ.
Q.6. Many enzymes require a non-protein component called an apoenzyme and a protein component called a co-factor to become active.
Q.7. Cadmium and beryllium are believed to inactivate enzymes by blocking the sites on the enzyme where such co-factors as iron normally attach.
Q.8. When lead covalently bonds to an enzyme, its inhibition of enzymes is considered to be irreversible.
Q.9. The exposure of allergens can trigger a diminished immune response in some people.
Q.10. Chemical pollutant such as ozone can depress the immune response by inactivating alveolar macrophages.
Q.11. B cells are the principle agents in cell-mediated immunity.
Q.12. Humoral immune responses are characterized by subcutaneous bleeding.
Q.13. The environmental pollutants such as ozone and fine particulates contribute to the significant rise in the numbers and severity of asthma cases.

Q.14. If absorbed, lead tends to be stored mostly in fatty tissue.
Q.15. Dioxin is considered to be one of the most toxic natural chemicals.
Q.16. The EPA has listed 20 μg/dL as the maximum acceptable blood lead level for fetuses and young children.
Q.17. Lead may impair fertility in both men and women when blood lead levels approach 50 μg/dL.
Q.18. The process by which plants turn carbon dioxide and water into nutrients is known as photosynthesis.
Q.19. Allergic contact dermatitis is a non-immune response caused by a direct action of an agent on the skin.
Q.20. The primary site of kidney damage resulting from acute exposure to inorganic mercury salts is the glomerulus.

Answers (Exercise 6)

1. F	2. F	3. F	4. T	5. F	6. F	7. F	8. T	9. F	10. T
11. F	12. T	13. T	14. F	15. F	16. F	17. T	18. T	19. F	20. F

Exercise 7

Q.1. If only 20% of the volume of petroleum is contained in pipes underground attached to the tank, the system will qualify as a UST under federal regulations.
Q.2. A company or agency remains responsible for the proper disposal of hazardous waste even after it leaves their property.
Q.3. In order for an employee to be prosecuted for negligent violation (giving rise to criminal liability), the law must demonstrate the actions were intentional.
Q.4. A large quantity generator of is defined as one that produces 1000 kg or more of hazardous wastes in a year.
Q.5. In some states, used oil (such as motor oil) that is not recycled is designated a hazardous waste.
Q.6. A waste is considered hazardous if it has a pH greater than 10.
Q.7. The first step in implementing CERCLA is to locate or find the hazardous waste site(s).
Q.8. The purpose of SARA Title III is to assure the public and emergency response agencies that information regarding hazardous chemicals is available to them.
Q.9. The Department of Transportation (DOT) regulates the disposal of hazardous wastes shipped by truck.
Q.10. Source reduction must be reported by facilities generating over a certain amount of toxic emissions during the previous calendar year.
Q.11. Septic tanks, heating oil tanks, and residential fuel tanks are regulated under the federal Underground Storage Tanks (UST) law.

Answers (Exercise 7)

1. F 2. T 3. F 4. F 5. T 6. F 7. T 8. T 9. F 10. T 11. F

16.3 FILL IN THE BLANKS

Exercise 8

Q.1. The process of contamination of environment (air, water, soil) through discharge of harmful substances is known as _______.
Q.2. The size of the particles that can reach alveoli of lungs is less than _______.
Q.3. The pollutants released directly into atmosphere either through natural or anthropogenic sources are called _______ pollutants.
Q.4. Pollutants formed due to the interaction of pollutants already present in the atmosphere are called _______ pollutants.
Q.5. Man-made sources or anthropogenic sources constitute _______ of pollution.
Q.6. The major air pollutant is _______.
Q.7. The pollutant that is produced due to incomplete combustion is _______.
Q.8. The lethal concentration of CO in air is _______.
Q.9. The major source of CO in urban setting is _______.
Q.10. CO binds with haemoglobin forming _______.
Q.11. The colour of blood in CO poisoning is _______.
Q.12. The main symptom in CO poisoning is _______.
Q.13. The specific therapy for CO poisoning is _______.
Q.14. The normal concentration of carbon dioxide in atmosphere is _______.
Q.15. Symptoms of CO_2 poisoning are evident at a concentration of _______ in air.
Q.16. The process of elevation of earth's temperature due to an increase in greenhouse gases in the atmosphere is called _______.
Q.17. Greenhouse gas of animal origin, which is implicated in global warming, is _______.
Q.18. The primary pollutants responsible for acid rains are _______.
Q.19. The level of SO_2 in air which is considered dangerous is _______.
Q.20. _______ gas is used as preservatives in canned meat products.

Answers (Exercise 8)

1. Pollution
2. 1 μm
3. Primary
4. Secondary
5. 98%
6. Carbon monoxide (CO)
7. CO
8. 400 PPM
9. Automobiles
10. Carboxy haemoglobin
11. Cherry red
12. Hypoxia
13. Oxygen
14. 0.5% (5,000 PPM)
15. 5%
16. Global warming
17. Methane gas
18. Sulphur oxides
19. 100 PPM
20. SO_2

Exercise 9

Q.1. 'Silo-fillers disease' or 'Silage gas' poisoning is caused by _______.
Q.2. The level of nitrogen oxides (NO_2 and N_2O_4) in air that can cause harmful effects is _______.
Q.3. The major system affected by NO_2 and SO_2 poisoning is _______ system.
Q.4. Painter's syndrome is caused by continuous exposure to _______ vapours.
Q.5. The main symptoms in chronic exposure to solvents is _______.
Q.6. The pollutant that can form ozone in the atmosphere by absorbing UV light is _______.
Q.7. The pollutant that can cause permanent damage to lungs even with short-term exposure in low concentrations is _______.
Q.8. The combination of smoke and fog results in the formation of _______ which considerably reduces visibility.
Q.9. The important contributor for development of photochemical smog is _______.
Q.10. The level of total dissolved solids (TDS) in drinking water that causes no hazard to animals is _______.
Q.11. The species of animal that is highly sensitive for polychlorinated biphenyls (PCB) toxicity is _______.
Q.12. PCB inhibits the synthesis of the central neurotransmitter in _______ brain.
Q.13. Feminization of the male fetus is caused by _______ pollutants.
Q.14. 'Cola-coloured babies' are born when mothers are exposed to _______ pollutants during pregnancy.
Q.15. 'Chick edema' is a characteristic clinical condition produced by _______ pollutants.
Q.16. Water-treatment processes such as chlorination leads to the production of _______ due to the reaction of chlorine with organic matter.

Answers (Exercise 9)

1. Nitrogen dioxide (NO_2)
2. 100 PPM
3. Respiratory
4. Solvent
5. Encephalopathy
6. NO_2
7. Ozone (O_3)
8. Smog
9. Peroxy acetyl nitrate (PAN)
10. <1000 PPM
11. Mink
12. Dopamine
13. PCB
14. PCB
15. PCB
16. Trihalomethanes (THM).

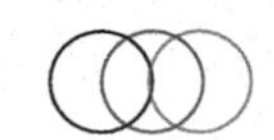

16.4 MATCH THE STATEMENTS

(Column A with Column B)

Exercise 10

Column A		Column B	
Q.1.	photo-degradation	a.	regulate body temperature
Q.2.	sunburn	b.	cell-mediated
Q.3.	biotransformation	c.	air mixture
Q.4.	pollutants	d.	radiation
Q.5.	aerosol	e.	benzo (*a*) pyrene
Q.6.	$PM_{2.5}$	f.	light
Q.7.	PAHs	g.	modification
Q.8.	skin	h.	micrometer
Q.9.	hypersensitivity reactions	i.	ozone
Q.10.	first trophic level	j.	plant

Answers (Exercise 10)

1. f 2. d 3. g 4. i 5. c 6. h 7. e 8. a 9. b 10. j

SECTION 8
APPLICATIONS IN TOXICOLOGY

Chapter 17

FORENSIC AND CLINICAL TOXICOLOGY

17.1 MULTIPLE-CHOICE QUESTIONS

(Choose the most appropriate response.)

Exercise 1

Q.1. Toxicity associated with any chemical substance is referred to as ________.
  a) poisoning
  b) intoxication
  c) over dosage
  d) toxicology

Q.2. Clinical toxicity, which one is secondary to accidental exposure ________.
  a) toxicology
  b) intoxication
  c) poisoning
  d) overdose

Q.3. Chest pain is related to ________.
  a) neurological examination
  b) cardiopulmonary examination
  c) gastrointestinal examination
  d) both cardiopulmonary examination and gastrointestinal examination

Q.4. Technique in which anticoagulated blood is passed through a column containing activated charcoal or resin particles is referred to as ________.
  a) whole bowel irrigation
  b) forced dieresis
  c) haemodialysis
  d) hemoperfusion

Q.5. Which of the following substances is *not* easily adsorbed by activated charcoal?
  a) Iron
  b) Ethanol
  c) Methanol
  d) All of the above

Q.6. The effect of syrup of ipecac starts within 30 minutes of administration and lasts for approximately ________.
  a) 30 minutes
  b) 1 hours
  c) 1 hours and 30 minutes
  d) 2 hours

Q.7. Which of the following procedures is contraindicated for patients who have ingested strong acids________?
  a) Emesis
  b) Gastric lavage
  c) Whole bowel irrigation
  d) Both emesis and gastric lavage

Q.8. Which of the following technique is helpful in removing ethanol from body?
a) Dialysis
b) Activated charcoal
c) Diuresis
d) Hemoperfusion

Q.9. The most effective treatment in GI decontamination with acetaminophen is ________.
a) emesis
b) gastric lavage
c) activated charcoal
d) dialysis

Q.10. Drug X is available as a 2.5% solution for intravenous administration. The desired dosage of this drug is 5 mg/kg. What volume of drug should be injected if the patient weighs 50 kg?
a) 0.2 mL
b) 1.0 mL
c) 2.0 mL
d) 10 mL
e) 20 mL

Q.11. Thalidomide was accidentally discovered as a ________.
a) cardiotoxic agent
b) liver tonic
c) sedative/tranquilizer
d) cough mixture

Q.12. Barbiturate, benzodiazepine abuse, dependency and sedative intoxication is generally associated with ________.
a) slurred speech
b) uncoordinated motor movements
c) impairment attention
d) all of the above

Q.13. A target organ of toxicity is ________.
a) lung
b) heart
c) reproductive system
d) kidney
e) liver

Q.14. A single large dose of *N*-nitrosodimethylamine fails to induce cancer in rats, but repeated dosing induces cancer because ________.
a) the single large dose is lethal, while the threshold for cancer induction can be exceeded by repeated smaller doses
b) the main DNA lesion from a single large dose can be repaired readily by methyl transferase, while repeated smaller doses can deplete the available repair enzyme, induce mutations in DNA, and effectively induce cancer
c) the enzyme system involved in detoxification of *N*-nitrosodimethylamine is depleted after repeated doses, allowing *N*-nitrosodimethylamine to build up and exceed the threshold for cancer induction
d) the enzyme system involved in conversion of *N*-nitrosodimethylamine to the active carcinogen is induced and on subsequent repeated doses more active carcinogen is produced
e) the initial dose of *N*-nitrosodimethylamine causes cell damage and, thus, high mitotic rates; subsequent small doses induce mutations in DNA and effectively induce cancer

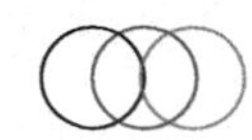

Q.15. Which of the following antidotes is *not* used in cyanide poisoning?
a) Dicobalt EDTA
b) Hydroxycobalamin
c) Sodium nitrite
d) Dimercaprol

Q.16. The duration of an ultrashort-acting barbiturate is ________.
a) 3 hours
b) 3 hours
c) 15–20 minutes
d) 0 minutes

Q.17. For each of the following substances, indicate which of the substances would result in increased androgen production in a male athlete ________.
a) cortisol
b) dehydroepiandrosterone
c) growth hormone
d) luteinizing hormone
e) salbutamol

Q.18. For each of the following drug or drug groups, which has been established for wrestling, boxing, or horse-riding use?
a) Diuretics
b) Beta blockers
c) AASs (androgenic-anabolic steroids)
d) Streptomycin

Q.19. For each of the following drug or drug groups, which has been established for precision sports, yachting, soccer, and modern pentathlon use?
a) Diuretics
b) Beta blockers
c) AASs (androgenic-anabolic steroids)
d) Streptomycin

Q.20. For each of the following drug or drug groups, which has been established for weightlifting, track and field, bodybuilders, and footballers use?
a) Diuretics
b) Beta blockers
c) AASs (androgenic-anabolic steroids)
d) Streptomycin

Q.21. Which of the following drugs does *not* cause a prolonged QRS?
a) Thioridazine
b) Propranolol
c) Quinine
d) Metoprolol

Q.22. Which of the following antidotes is *not* used in cyanide poisoning?
a) Dicobalt EDTA
b) Hydroxycobalamin
c) Sodium nitrite
d) Dimercaprol

Q.23. Regarding 'Tests for Drugs' in toxicology, which statement is *false*?
a) Bedside ECG and serum-paracetamol are regarded as routine toxicology screening tests.
b) Fluorescence polarization immunoassay on urine or blood samples is used for 'drug screening.'

c) Gas chromatography/mass spectrometry is performed as a 'confirmatory test' on blood or urine samples.
d) Thin layer/paper chromatography used on urine and blood samples assists in 'drug screening.'

Q.24. Extracorporeal elimination of drugs may be of use in all of the following *except* ________.
a) ethylene glycol
b) salicylates
c) atenolol
d) organophosphates

Q.25. The following statements about Digibind are true *except* ________.
a) indicated when there is digoxin ingestion of >10 mg
b) 40 mg binds approximately 0.6 mg digoxin
c) serum digoxin levels increase following its administration
d) indicated for use if serum digoxin level is >10 nmol/L in acute overdose

Q.26. The following is contraindicated to treat theophylline seizures:
a) Diazepam
b) Phenobarbitone
c) Chloral Hydrate
d) Phenytoin

Q.27. Following an aspirin overdose, the initial acid-base derangement is usually
a) Respiratory acidosis
b) Metabolic acidosis
c) Respiratory alkalosis
d) Metabolic alkalosis

Q.28. Which of the following pairs is *false* regarding drugs and their appropriate antidotes?
a) Beta blockers – glucagon
b) Chloroquine – Diazepam
c) Isoniazid – Pralidoxime
d) Methanol – ethanol

Q.29. With regard to sympathomimetic toxicity, which of the following is correct?
a) Co-ingestion of cocaine and alcohol results in greater neurological toxicity than cocaine alone.
b) There is no difference between intravenous or oral amphetamine use and the incidence of rhabdomyolysis.
c) Patients with psychomotor acceleration and psychosis should initially be reviewed by the psychiatric team.
d) Auditory hallucinations are uncommon.

Q.30. The maximum safe dose for paracetamol every 24 hours is ________.
a) 90 mg/kg in children
b) 150 mg/kg in children
c) 200 mg/kg in children
d) in an adult up to 5 g

Q.31. Theophylline toxicity ________.
a) often presents with abdominal pain, hematemesis, and drowsiness
b) causes its effects by blockade of voltage-sensitive calcium channels in cardiac muscle and CNS
c) is rarely fatal with good supportive care
d) may cause refractory seizures

Q.32. Which statement is *false* regarding antihistamine toxicity?
a) Doxylamine overdose may result in non-traumatic rhabdomyolysis.
b) The first generation antihistamines are a common cause for patients presenting with anticholinergic toxicity in ED.

 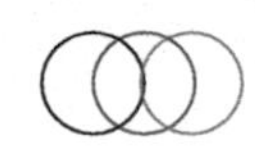

c) Phenytoin is indicated for managing seizures.
d) Diphenhydramine and dimenhydrinate may cause cardiac conduction delays similar to tricyclic antidepressant overdose.

Q.33. Which statement is *false* regarding colchicine poisoning?
a) Colchicine is rapidly absorbed following oral administration.
b) The multiorgan failure phase typically occurs 24 hours after ingestion.
c) A rebound leukocytosis occurs 3 weeks after poisoning in survivors, signalling recovery of bone marrow function.
d) Charcoal is indicated for gut decontamination.

Q.34. Which statement is true regarding anticonvulsant drug poisoning?
a) Chronic toxicity with therapeutic dosing is uncommon with Phenytoin.
b) A poisoning with Sodium Valproate at 100 mg/kg is likely to result in coma.
c) Cardiac monitoring is not required where Phenytoin is the only agent ingested.
d) Carbamazepine levels are not useful in the management of carbamazepine poisoning.

Q.35. Regarding antimicrobial toxicity, the following are often fatal *except*:
a) Isoniazid
b) Neomycin
c) Chloroquine
d) Quinine

Q.36. Regarding Isoniazid toxicity, all of the following are true *except* ________.
a) Metabolic acidosis is common.
b) Treatment of seizures is best treated with high-dose BDZ.
c) Acidosis is thought to be secondary to seizures.
d) Toxicity is seen early post-ingestion (within 1–2 hours).

Q.37. Which is *false* regarding Quinine poisoning:
a) Deliberate overdose is often fatal.
b) Significant overdose may result in cardiovascular collapse.
c) Deliberate overdose may result in permanent blindness.
d) PR interval prolongation is a major change seen in the ECG.

Q.38. In an overdose, the following are true *except* ________.
a) Penicillins and cephalosporins are associated with seizures.
b) Seizures in isoniazid toxicity are due to pyridoxine disruption.
c) A farmer presenting with hallucinations, dementia, and exquisitely painful legs may well have ergot poisoning.
d) A patient presenting after ingestion of colchicine presenting with minimal symptoms can be safely discharged after a short period of observation.

Q.39. In a patient with a history of unknown psychiatric medication, which of the following is true?
a) Hyperreflexia, rigidity, and hyperthermia would likely represent a dose-related effect of olanzapine.
b) Extrapyramidal effects make it more likely to be a typical than atypical antipsychotic.
c) Positive ECG changes in TCA toxicity are highly predictive of likely arrhythmias.
d) If you suspect serotonin syndrome in an intubated patient, then midazolam and fentanyl sedation would be a good choice due to its short duration of action.

Q.40. Which of the following is *least* likely to be helpful in a calcium channel blocker overdose?
a) Atropine
b) Intra-aortic balloon counterpulsation
c) Insulin
d) Resonium

Q.41. Concerning morphine:
a) An active metabolite of hydromorphone
b) Readily extracted from strong alkaline solution

c) Urinary metabolites include morphine-glucuronide
d) Biotransformed to 6-acetylmorphine
e) Readily extracted from strong acid solution

Q.42. Oxazepam is a metabolite of which of the following?
a) Diazepam
b) Alprazolam
c) Lorazepam
d) Flurazepam
e) Flunitrazepam

Q.43. In gas chromatography, which of the following has the longest retention time on a 50% phenylmethyl or HP-17 liquid phase?
a) Nicotine
b) Meperidine
c) Strychnine
d) Diazepam
e) Phentermine

Q.44. A specimen of known concentration used to verify a calibration is ________.
a) calibrator
b) control
c) reference
d) standard
e) blank

Q.45. A 200-pound man consumes 6 × 12-oz. beers and 2 × 1-oz. shots of whiskey (100 proof) between 9:00 PM and 11:00 PM. A breath alcohol test performed at 1:00 AM would be expected to give an alcohol concentration in the following range (g/210 L):
a) 0.04–0.06
b) 0.07–0.09
c) 0.13–0.15
d) 0.16–0.18
e) 0.10–0.12

Answers (Exercise 1)

1. b	6. d	11. c	16. c	21. d	26. d	31. d	36. b	41. c
2. c	7. a & b	12. d	17. b & d	22. d	27. c	32. c	37. d	42. a
3. b	8. a	13.d	18. a	23. d	28. c	33. c	38. d	43. c
4. d	9. c	14. b	19. b	24. d	29. d	34. c	39. b	44. b
5. d	10. d	15. d	20. c	25. d	30. a	35. b	40. d	45. e

Exercise 2

Q.1. Which of the following is most commonly used as a drug of sexual assault?
a) Narcotics
b) Amphetamines
c) Benzodiazepines
d) Ethanol
e) Antidepressants

Q.2. All of the following statements regarding analytic/forensic toxicology are true *except* ________.
a) Analytic toxicology uses analytic chemistry to characterize a chemical's adverse effect on an organism.
b) Medical examiners and coroners are most important in determining cause of death.
c) Tissues and body fluids are vital in forensic toxicology.

 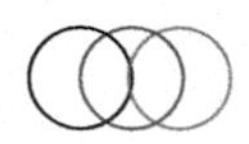

d) Forensic toxicology is used for purposes of the law.
e) Chapuis first characterized a system for classifying toxic agents.

Q.3. Which of the following criteria is *not* routinely used to check for the adulteration of a drug urine analysis?
a) Urea
b) pH
c) Colour
d) Specific gravity
e) Creatinine

Q.4. Which blood alcohol concentration (BAC) is most commonly used as the statutory definition of DUI (driving under influence)?
a) 0.04
b) 0.06
c) 0.08
d) 0.12
e) 0.16

Q.5. Which of the following drugs is *not* properly matched with its most common analytic method?
a) Benzodiazepines – GC/MS
b) Ibuprofen – LC/HPLC
c) Amphetamines – immunoassays
d) Barbiturates – GC/immunoassays
e) Ethanol – immunoassays

Q.6. For which of the following drugs is serum *not* used during toxicology testing?
a) Ethanol
b) Cocaine
c) Aspirin
d) Barbiturates
e) Ibuprofen

Q.7. Which of the following is *least* important in determining variability in response to drug therapy?
a) Drug interactions
b) Distribution in body tissue
c) Body mass index
d) Pathologic conditions
e) Rate of metabolism

Q.8. Which of the following statements is *false* regarding steady state?
a) Steady-state concentrations are proportional to the dose/dosage interval.
b) Steady state is attained after approximately four half-lives.
c) The steady-state concentrations are proportional to F/Cl.
d) Monitoring of steady-state drug concentration assumes that an effective concentration is present.
e) Fluctuations in concentration are increased by slow drug absorption.

Q.9. Which of the following is an indirect method of measuring a chemical for its metabolite?
a) Blood test
b) Hair sample
c) Urinalysis
d) Haemoglobin adduct detection
e) Breath analysis

Q.10. Which of the following statements regarding analytic/forensic toxicology is true?
a) Antidepressants are commonly used to incapacitate victims.
b) It is easy to test for and prove that marijuana is a factor in an automobile accident.

c) Heroin is the drug most commonly encountered in emergency toxicology.
d) Toxicologists can play an important role in courtroom testimonies.
e) Ethanol intoxication often results in death.

Q.11. What is the primary goal in taking a history in a poisoned patient?
a) Determining drug allergies
b) Determining susceptibility to drug overdose
c) Determining likelihood of an attempted suicide
d) Determining the ingested substance
e) Determining the motive behind the poisoning

Q.12. Who is most likely to give incorrect information while taking a history of a poisoned patient?
a) Patient
b) Emergency medical technician (EMT)
c) Employer
d) Pharmacist
e) Family members

Q.13. Which of the following sets of clinical features characterizes an anticholinergic toxic syndrome?
a) Increased blood pressure, decreased heart rate, decreased temperature
b) Decreased blood pressure, increased heart rate, decreased temperature
c) Increased blood pressure, increased heart rate, increased temperature
d) Decreased blood pressure, decreased heart rate, decreased temperature
e) Increased blood pressure, decreased heart rate, increased temperature

Q.14. Which of the following sets of clinical features characterizes a sympathomimetic toxic syndrome?
a) Miosis, decreased bowel sounds, decreased alertness
b) Decreased heart rate, increased temperature, mydriasis
c) Hyperalertness, decreased blood pressure, miosis
d) Increased temperature, increased heart rate, miosis
e) Mydriasis, increased blood pressure, hyperalertness

Q.15. Which of the following drugs *cannot* be tested for in a hospital on a stat basis?
a) Ethanol
b) Cocaine
c) Aspirin
d) Phenytoin
e) Digoxin

Q.16. Which is *not* included in the differential diagnosis of an elevated anion gap?
a) Ethanol
b) Methanol
c) Diabetes
d) Ethylene glycol
e) Diarrhoea

Q.17. An elevated osmol gap might suggest which of the following?
a) Methanol poisoning
b) Chronic vomiting
c) Lactic acidosis
d) Diabetic ketoacidosis
e) Chronic diarrhoea

Q.18. Which of the following is *least* likely to prevent further poison absorption?
a) Induction of emesis
b) Activated charcoal

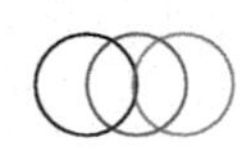

c) Gastric lavage
d) Syrup of ipecac
e) Parasympathetic agonist

Q.19. Which of the following would *not* be used to enhance poison elimination?
a) Oral activated charcoal
b) Hemoperfusion
c) Acidification of urine
d) Hemodialysis
e) Plasma exchange

Q.20. Which of the following might be used as an antidote for patients with cyanide poisoning?
a) Syrup of ipecac
b) Atropine
c) Chelating agents
d) Sodium nitrite
e) Quinine

Q.21. Administration by oral gavage of a test compound that is highly metabolized by the liver versus subcutaneous injection will most likely result in ________.
a) less parent compound present in the systemic circulation
b) more local irritation at the site of administration caused by the compound
c) lower levels of metabolites in the systemic circulation
d) more systemic toxicity
e) less systemic toxicity

Q.22. Intoxication from consumption of wild cherry or apricot pits would best be treated by ________.
a) hyperbaric oxygen
b) artificial respiration
c) inhalation of amyl nitrite
d) intravenous sodium nitrite and sodium thiosulfate
e) oral sodium nitrate

Q.23. The most useful bedside test to suggest snake bite envenomation is ________.
a) prothrombin time
b) Q.20-min whole-blood clotting time
c) international normalized ratio
d) platelet count

Q.24. A 12-year-old boy had an alleged history of snake bite and presented to the hospital with inability to open eyes well and difficulty in breathing. He is very anxious and is having tachycardia and tachypnea. On examination, the bite mark cannot be visualized, and there is no swelling of the limb. He has bilateral ptosis. His 20-min whole-blood clotting test is good quality. What is the next course of action?
a) Don't give anti-snake venom (ASV), but observe the patient
b) Give ASV and keep the patient in observation
c) Give ASV, and give neostigmine and observe the patient
d) Reassure the patient and send him home with anxiolytic

Q.25. Magnan's symptoms are characteristic symptoms with which poisoning?
a) Alcohol
b) Charas
c) Cocaine
d) Ecstasy

Q.26. Which of the following is the correct xenobiotic:toxicity pair?
a) Adriamycin: tubular proteinuria
b) Polyvinyl alcohol:disseminated intravascular coagulation

c) Hydralazine:renal papillary necrosis
d) Maleic acid:mesangial fibrosis
e) Gold:glomerulonephritis

Q.27. The primary site of kidney damage resulting from acute exposure to inorganic mercury salts is the ________.
a) glomerulus
b) proximal tubule
c) loop of Henle
d) renal papilla
e) entire nephron

Q.28. Which substance has been identified as a respiratory tract carcinogen in humans?
a) Kaolin
b) Hydrogen fluoride
c) Arsenic
d) Cotton dust
e) Vanadium

Q.29. Chloracne is associated with ________.
a) prominent hyperkeratosis of the follicular canal
b) production of excessive sebum
c) exposure to halogenated aliphatic hydrocarbons
d) exposure to chlorine gas
e) increases in serum androgen levels

Q.30. Toxic injury to the cell body, axon, and surrounding Schwann cells of peripheral nerves are referred to, respectively, as ________.
a) neuropathy, axonopathy, and myelopathy
b) neuronopathy, axonopathy, and myelinopathy
c) neuropathy, axonopathy, and gliosis
d) neuronopathy, dying-back neuropathy, and myelopathy
e) chromatolysis, axonopathy, and glia cells

Q.31. A person was brought by police from the railway platform. He is talking irrelevantly. He is having dry mouth with hot skin, dilated pupils, staggering gait, and slurred speech. The most probable diagnosis is ________.
a) alcohol intoxication
b) carbamates poisoning
c) organophosphorus poisoning
d) datura poisoning

Q.32. Hyperthermia in a patient of poisoning is a pointer to all *except* ________.
a) ecstasy
b) selective serotonin reuptake inhibitor
c) salicylates
d) chlorpromazine

Q.33. In methyl alcohol poisoning, there is central nervous system depression, cardiac depression, and optic nerve atrophy. These effects are produced due to ________.
a) formaldehyde and formic acid
b) acetaldehyde
c) pyridine
d) acetic acid

Q.34. A 39-year-old carpenter has taken two bottles of liquor from the local shop. After about an hour, he develops confusion, vomiting, and blurring of vision. He has been brought to the emergency department. He should be given ________.
a) naloxone
b) diazepam

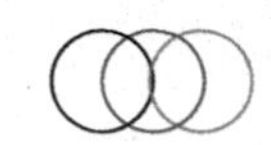

c) flumazenil
d) ethyl alcohol

Q.35. Phosphine liberated in the stomach in aluminum phosphide poisoning is toxic to all *except* ________.
a) lungs
b) kidneys
c) liver
d) heart

Q.36. Paraquat poisoning causes ________.
a) renal failure
b) cardiac failure
c) respiratory failure
d) multiple organ failure

Q.37. Ecstasy toxicity causes ________.
a) hyperreflexia
b) trismus
c) dilated pupils
d) visual hallucinations
e) all of the above

Q.38. A housewife ingests a rodenticide white powder accidentally. She is brought to the hospital where the examination shows generalized, flaccid paralysis and an irregular pulse. electrocardiogram shows multiple ventricular ectopics, generalized changes with ST-T. Serum potassium is 2.5 mEq/L. The most likely ingested poison is ________.
a) barium carbonate
b) super warfarins
c) zinc phosphide
d) aluminum phosphide

Q.39. All of the following are treatment options for toxic alcohol poisoning *except* ________.
a) fomepizole
b) hydroxycobalamin
c) thiamine
d) folic acid
e) Pyridoxine

Answers (Exercise 2)

1. d	6. b	11. d	16. e	21. a	26. e	31. d	36. d
2. b	7. c	12. a	17. a	22. d	27. b	32. d	37. e
3. a	8. e	13. c	18. e	23. b	28. c	33. a	38. a
4. c	9. d	14. e	19. c	24. c	29. a	34. d	39. b
5. e	10. d	15. b	20. d	25. c	30. b	35. b	

Exercise 3

Q.1. Digoxin immune Fab therapy is *not* indicated in which of the following natural teas/broths?
a) Oleander
b) Lily of the valley
c) Cane toad
d) Sea horse

Q.2. With respect to Theophylline toxicity, which is *false*?
  a) Anxiety, vomiting, and tremor are early manifestations.
  b) It can precipitate supraventricular tachycardia.
  c) Hypoglycemia, hypophosphatemia, and hypomagnesemia are complications.
  d) Beta blockers are contraindicated.

Q.3. Which of the following paired agents or syndrome/interventions is *false*?
  a) Anticholinergic agents:physostigmine
  b) Neuroleptic malignant syndrome:bromocryptine
  c) Serotonin syndrome:cyproheptadine
  d) Organophosphates:atropine, pyridoxine

Q.4. Regarding clozapine overdose, which one is true?
  a) Ingestion of a single tablet in a child needs assessment in hospital.
  b) Acute poisoning is associated with agranulocytosis.
  c) Patients typically become comatose and require endotracheal intubation with significant overdose.
  d) Overdose is not associated with anticholinergic effects.

Q.5. Which of the following statements is *false* regarding iron overdose?
  a) A serum iron level should be done 12 hours following ingestion.
  b) An anion-gap metabolic acidosis is typical.
  c) Activated charcoal is not indicated because it does not adsorb iron.
  d) Abdominal X-ray may be useful.

Q.6. With regard to an iron overdose, which statement is *false*?
  a) Accidental childhood ingestion is usually not serious.
  b) Desferrioxamine is indicated if there are signs of systemic toxicity or a 4–6 hour level of greater than 90 μmol/L.
  c) Whole-bowel irrigation is recommended for ingestions of greater than 60 mg/kg confirmed on X-ray.
  d) Significant toxicity causes normal anion-gap acidosis.

Q.7. Which of the following significant toxic ingestions would *not* require early activated charcoal to ensure a good outcome?
  a) Paraquat > 50 mg/kg
  b) Sodium Valproate > 1 g/kg
  c) Colchicine > 0.8 mg/kg
  d) Bupropion > 9 g

Q.8. Common causes of toxic seizures in Australia include all *except* ________.
  a) alcohol
  b) tramadol
  c) venlafaxine
  d) bupropion

Q.9. A drug that can mimic brain death when taken in overdose is ________.
  a) thiopentone
  b) propranolol
  c) quetiapine
  d) baclofen

Q.10. Which of the following statements regarding clozapine overdose is true?
  a) A child that ingested a single tablet needs to be assessed and observed in the hospital.
  b) Acute poisoning is associated with agranulocytosis.
  c) Patients typically become comatose and require endotracheal intubation with a significant overdose.
  d) Patients need to be observed in the hospital for at least 24 hours.

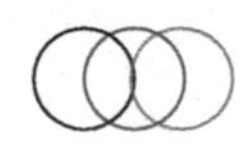

Q.11. Which of the following statements is *false* regarding iron overdose?
 a) A serum iron level should be done 12 hours following ingestion.
 b) An anion-gap metabolic acidosis is typical.
 c) Activated charcoal is not indicated because it does not adsorb iron.
 d) Hypoglycemia is a rare feature of severe iron poisoning.

Q.12. Which of the following is *not* a benign presentation?
 a) A child with a normal GIT who bites a mercury thermometer and swallows some mercury.
 b) A child who ingests 30 mg/kg of elemental iron.
 c) A child who ingests one of her brother's Risperidone tablets.
 d) A child who ingests 1 g metformin.

Q.13. Which pairing is *incorrect*?
 a) Lead encephalopathy–sodium calcium edetate
 b) Isoniazid overdose–pyridoxine
 c) Methemoglobinemia–methylene blue
 d) Methotrexate overdose–cyproheptadine

Q.14. Amisulpride overdose commonly results in ________.
 a) abrupt cardiovascular collapse up to 12 hours post ingestion with large overdoses
 b) torsades at doses of 2–4 g
 c) serotonin syndrome if co-ingested with a serotonergic compound
 d) seizures with massive overdose

Q.15. Which of the following statements regarding sympathomimetic toxicity is *incorrect*?
 a) MDMA may cause SIADH.
 b) Cocaine is a sodium channel blocker.
 c) Lignocaine is used as a second-line agent to control ventricular dysrhythmias in cocaine overdose.
 d) Beta blockers are used as a first-line agent to control hypertension and tachycardia in a methamphetamine overdose.

Q.16. Which of the following is true regarding β-blocker toxicity?
 a) Propranolol facilitates sodium entry into cells resulting in QT prolongation on ECG.
 b) Significant toxicity is usually apparent within 6 hours.
 c) Sotalol is the only β-blocker that causes QT prolongation.
 d) β-blocker overdose causes decreased intracellular cAMP concentrations.

Q.17. Regarding calcium channel blockers (CCB):
 a) Nifedipine produces more cardiotoxic effects than verapamil.
 b) Activated charcoal therapy is not indicated in CCB overdose.
 c) Calcium and glucagon form the mainstay of treatment in CCB overdose.
 d) Standard CCB preparations are rapidly absorbed from the GI tract with onset of action within 30 minutes.

Q.18. Regarding Warfarin, which is *false*?
 a) Activated charcoal may be useful for decontamination.
 b) A normal INR at 48 hours excludes an acute Warfarin ingestion.
 c) Prothrombinex contains Factors II, IX, and X.
 d) Super Warfarin ingestion is rarely fatal.

Q.19. Regarding β-blocker toxicity, which is *false*?
 a) Activated charcoal may be useful for decontamination.
 b) Sotalol is associated with prolonged QRS.
 c) Seizures are seen in propranolol poisoning.
 d) PR prolongation is an early sign of toxicity.

Q.20. Which medication is *not* implicated in serotonin syndrome?
 a) Fentanyl
 b) Ondansetron

c) Valproic acid
d) Mirtazapine

Q.21. Which one of the following is true concerning salicylate intoxication?
a) High blood levels cannot be removed by dialysis.
b) If a respiratory alkalosis is present, do not administer intravenous bicarbonate.
c) Salicylate intoxication causes both a metabolic acidosis and a metabolic alkalosis.
d) The recommended treatment is intravenous fluids without dextrose.
e) Oil of wintergreen can cause salicylate poisoning.

Q.22. A 35-year-old woman presents with an acute lithium overdose. Which one of the following statements concerning lithium is true?
a) Aggressive diuresis is needed to augment lithium excretion.
b) Hypocalcemia can be seen as a side effect of lithium.
c) Lithium cannot be removed by dialysis.
d) It is recommended that you avoid the use of saline in lithium intoxication.
e) You should evaluate thyroid function in anyone taking lithium.

Q.23. Which one of the following is the treatment for a heparin overdose?
a) Vitamin K
b) Fresh frozen plasma
c) Protamine sulfate
d) Desmopressin acetate (DDAVP)
e) Cryoprecipitate

Q.24. In the case history of the young woman who was finally diagnosed with parasitic diseases, what was the screening test used to finally discover her illness?
a) Pap smear
b) Cardiography
c) Rectal smear
d) X-ray

Q.25. Which one of the following statements concerning digoxin is true?
a) Digoxin is used in treating diastolic heart failure.
b) Digoxin toxicity is treated with dialysis.
c) Digoxin dosing must be increased when kidney disease is present.
d) Amiodarone and quinidine can decrease digoxin levels.
e) Hypokalemia can exacerbate digoxin toxicity.

Q.26. A patient had a fever and altered mental status. He was recently started on fluphenazine (Prolixin). He is agitated and his temperature is 39.4°C (103°F). His blood pressure is 160/100 mmHg. A CPK level is 50,000. What is the most appropriate treatment at this time?
a) Urgent hemodialysis
b) Intravenous saline alone for the rhabdomyolysis
c) Lorazepam (Ativan) for agitation
d) Dantrolene
e) Cyproheptadine

Q.27. Humans have died from an immune reaction referred to as ________ caused by fly bites.
a) diphtheria toxic shock
b) heart inflammation
c) typhoid fever
d) anaphylactic shock

Q.28. All of the following symptoms are commonly found in sympathomimetic intoxications *except* ________.
a) agitation
b) tachycardia and hypertension
c) hyperthermia

 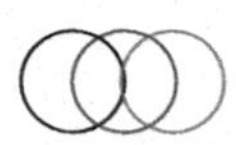

d) dry, flushed skin
e) mydriasis

Q.29. All of the following are effectively bound to activated charcoal *except* ________.
a) acetaminophen
b) tricyclic antidepressant
c) iron
d) theophylline
e) salicylates

Q.30. What is the most important therapy in the management of a serious benzodiazepine ingestion?
a) multidose activated charcoal
b) alkaline diuresis
c) aggressive airway management
d) cardiac monitoring
e) hemodialysis

Q.31. Which of the following is *not* a side effect of Digoxin toxicity?
a) Bradycardia
b) Yellow vision changes
c) Scooping of the T segment on an ECG
d) Hypokalemia
e) Gynecomastia

Q.32. Which of the following chelating agents is recommended for acute lead poisoning with signs of encephalopathy?
a) Succimer
b) Penicillamine
c) Dimercaprol
d) Calcium EDTA
e) Dimercaprol + calcium EDTA

Q.33. Which of the following dermatologic findings and potential causes is incorrect?
a) Cyanosis – methemoglobinemia
b) Erythroderma – boric acid
c) Pallor – carbon monoxide
d) Jaundice – hypercarotenemia (excess carrot intake)
e) Brightly flushed skin – niacin

Q.34. All of the following symptoms can occur with Ciguatera poisoning *except* ________.
a) myalgia
b) flushing
c) metallic taste
d) reversal of temperature sensation
e) sensation of loose, painful teeth

Q.35. Which of the following is with regard to acetaminophen toxicity?
a) The Rumack–Matthew nomogram may be used for both acute and chronic ingestions.
b) The APAP level should ideally be checked within 1–4 hours of ingestion.
c) The Rumack–Matthew nomogram applies for ingestions up to 48 hours post-ingestion.
d) *N*-Acetylcysteine (NAC) should be started within 8 hours of ingestion if an APAP level cannot be obtained.
e) Activated charcoal should be used for all sustained-release ingestions.

Q.36. Which of the following statements concerning acetaminophen toxicity is *false*?
a) Hepatotoxicity occurs because of a depletion of glutathione.
b) Drugs that enhance the cytochrome P450 system diminish the toxic potential.
c) Signs of hepatotoxicity do not occur until at 8 hours post ingestion.

d) Hepatic necrosis is centrilobular in distribution.
e) Cimetidine may be protective because of its ability to diminish hepatic metabolism.

Q.37. Ophotoxemia refers to ________.
a) organophosphorus poisoning
b) heavy metal poisoning
c) scorpion venom poisoning
d) snake venom poisoning

Q.38. Elapidaes are ________.
a) vasculotoxic
b) neurotoxic
c) musculotoxic
d) non-toxic

Q.39. All of the following are treatment options for toxic alcohol poisoning *except* ________.
a) fomepizole
b) hydroxocobalamin
c) thiamine
d) folic acid
e) pyridoxine

Q.40. Which is *false* regarding Paraquat poisoning?
a) Supplemental oxygen should be avoided.
b) It is associated with 'Paraquat tongue'.
c) A raised creatinine carries a poor prognosis.
d) Paraquat has an effect on the neuromuscular junction.

Q.41. Which of the following drug:antidote pairs are *least* appropriate in the ED?
a) Hydrofluoric acid burn:calcium chloride
b) Cyanide poisoning:thiosulfate
c) Clonidine overdose: Naloxone
d) Benzodiazepine overdose: Flumazenil

Q.42. Which statement is *false*?
a) In anticholinergic syndrome, death may result from hyperthermia and dysrhythmias.
b) Oil of wintergreen ingestion is associated with altered mental state, respiratory alkalosis, metabolic acidosis, and tinnitus.
c) Serotonin syndrome is associated with ocular myoclonus and hyperreflexia.
d) Cyproheptadine may be of benefit in neuroleptic malignant syndrome.

Q.43. Which statement is *false* in regard to lithium poisoning?
a) Peak serum levels occur within 2–4 hours of oral ingestion.
b) Significant ECG changes only occur at very high serum levels.
c) Clinical features of lithium toxicity can be observed even when serum levels are in the normal range.
d) Neurological features dominate the clinical presentation.

Q.44. Which statement is *false* in regard to theophylline poisoning?
a) The precise mechanism of toxicity is unknown.
b) Serum levels will confirm poisoning and are invaluable in ongoing management.
c) Seizures refractory to benzodiazepines should be treated with second-line agents including phenytoin and phenobarbitone.
d) Poisoning associated with chronic use is more common than acute ingestions.

Q.45. Which of the following is *not* a feature of pure fast sodium channel blockade?
a) QRS widening
b) Tachycardia
c) VT
d) VF

 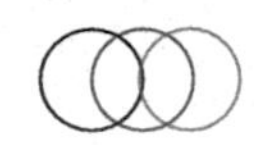

Q.46. In regards to β-blockers, which is *false*?
  a) Poisoning with most β-blockers is usually benign.
  b) Insulin:dextrose therapy may have a role.
  c) $NaHCO_3$ has an occasional role.
  d) Most symptomatic overdoses exhibit bradycardia.

Q.47. Regarding activated charcoal, which is *incorrect*?
  a) The enormous surface area provided by these particles of charcoal irreversibly absorbs most ingested toxins preventing further absorption from the GI tract.
  b) The major risk is charcoal pulmonary aspiration due to loss of airway reflexes associated with impaired consciousness or seizures.
  c) Ileus is not a contraindication to single-dose activated charcoal.
  d) Multiple-dose activated charcoal has the potential to enhance drug elimination by interruption to enterohepatic circulation and gastrointestinal dialysis.
  e) There is no data to support the use of activated charcoal in sorbitol or other cathartic agent over activated charcoal in water.

Q.48. Regarding urinary alkalinization, which is *incorrect*?
  a) Production of alkaline urine prevents reabsorption across the renal tubular epithelium, thus promoting excretion in the urine.
  b) In salicylate overdose, metabolism is saturated and the elimination half-life greatly prolonged.
  c) Severe established salicylate toxicity warrants a trial of urinary alkalinization rather than immediate hemodialysis.
  d) In phenobarbitone coma, multidose activated charcoal is superior to urinary alkalinization as the first line.

Q.49. Which of the following statements is *false* with regards to the toxicokinetics of phenytoin?
  a) HONK is a recognized complication.
  b) It is a Na channel blocker.
  c) It causes QRS widening.
  d) It shares the same order of elimination kinetics as salicylate.

Q.50. Induction of emesis is recommended as a detoxification procedure in dogs ingesting any of the following *except* ________.
  a) antifreeze (ethylene glycol)
  b) acetaminophen
  c) kerosene
  d) liquid aspirin
  e) chocolate

Q.51. The antidotal agent *N*-acetylcysteine is indicated for treatment of poisoning with ________.
  a) cholecalciferol rodenticides
  b) acetaminophen
  c) brodifacoum
  d) chlorpyrifos
  e) copper

Q.52. Which one of the following antidotes matches the underlying toxicity?
  a) Benzodiazepines–naloxone (Narcan)
  b) Narcotics–flumazenil (Romazicon)
  c) Ethylene glycol–ethanol (booze)
  d) Acetaminophen–fomepizole (4-methylpyrazole)
  e) High carboxyhemoglobin–methylene blue

Q.53. The toxicity produced by excessive ingestion of urea in cattle is due to ________.
  a) urea itself
  b) ammonia
  c) urea as well as ammonia

Q.54. Which of the following might be linked to parkinsonism?
a) Nitrogen dioxide
b) Zinc
c) Copper
d) Magnesium
e) Carbon monoxide

Q.55. Strychnine (gopher bait) causes arched back and neck (opisthotonos) and convulsions. What is the mechanism of action?
a) Uncoupled oxidative phosphorylation
b) Inhibits GABA receptors
c) Inhibits Na^+ influx and K^+ efflux
d) Inhibits acetylcholinesterase (AChE)
e) Inhibits glycine on anterior horn of spinal cord and endogenous transmitter release at Renshaw cells.

Q.56. Which animal will accumulate toxic molecules upon ingesting acetaminophen that result in methemoglobinemia and cell death?
a) Dog
b) Horse
c) Cat
d) Cow
e) Chicken

Q.57. A dog presents at the clinic with a wobbling, ataxic gait and drunken disposition. One of the main concerns is acute renal failure and life-threatening acidosis. What toxin do you think this animal ingested, and what is the treatment?
a) NSAID, treat gastric lavage
b) Acetaminophen, treat with *N*-acetyl cysteine
c) Ethylene glycol, treat with 4-MP
d) Ethylene glycol, treat with alcohol dehydrogenase
e) NSAID, treat with 4-MP

Q.58. Which clinical sign in small animals is *not* caused by chocolate?
a) Liver necrosis
b) CNS stimulation (seizure)
c) Tachycardia
d) Vasoconstriction
e) Vomiting

Q.59. Drug X is available as a 2.5% solution (w/v) in a 100-mL bottle. The desired intravenous dosage of this drug is 5 mg/kg body weight. What volume of drug should be injected if the patient weighs 50 kg?
a) 0.2 mL
b) 1.0 mL
c) 2.0 mL
d) 10 mL

Q.60. Ethanol, retinoids, valproic acid, and the angiotensin converting enzyme (ACE) inhibitors share what primary characteristic?
a) Induces liver toxicity
b) Lowers blood pressure
c) Causes developmental toxicity
d) Induces central nervous system effects

Q.61. A patient was admitted to the emergency room with decreased heart rate, blood pressure, body temperature, and pupil size. At intake, the patient appeared very sleepy, and a family member stated that he was recovering from recent surgery. Exposure to what class of agent explains the patient's symptoms?

a) A narcotic analgesic
b) A non-narcotic analgesic
c) An antacid
d) A benzodiazepine tranquilizer

Q.62. The antifungal drug ketoconazole specifically inhibits a cytochrome P450 isozyme responsible for the biotransformation of many drugs, among them midazolam. In a patient taking ketoconazole (200 mg/day PO), what would be the difference, if any, in this patient's exposure (AUC0-24h) to midazolam if midazolam was administered intravenously (IV) vs. orally?
a) No difference in IV vs. oral exposure because oral bioavailability of midazolam is 100% and is not metabolized by enterocytes
b) Increased exposure by the oral route because ketoconazole inhibits both the intestinal and hepatic p450 isozymes
c) Decreased exposure by the oral route because ketoconazole increases midazolam export from enterocytes via p-glycoprotein.
d) Decreased exposure by the IV route because ketoconazole preferentially inhibits p450 metabolism in enterocytes

Answers (Exercise 3)

1. d	6. d	11. a	16. d	21. e	26. d	31. d	36. b	41. d	46. d	51. b	56. c	61. a
2. d	7. b	12. c	17. b	22. e	27. d	32. e	37. d	42. d	47. a	52. c	57. c	62. b
3. d	8. a	13. d	18. c	23. c	28. d	33. c	38. b	43. b	48. c	53. b	58. a	
4. a	9. d	14. a	19. b	24. c	29. c	34. b	39. b	44. c	49. c	54. e	59. d	
5. a	10. a	15. b	20. d	25. e	30. c	35. d	40. d	45. b	50. c	55. e	60. c	

Exercise 4

Q.1. The most common pattern fingerprint is ________.
a) whorls
b) loops
c) composite
d) arches

Q.2. The pH of seminal fluid is ________.
a) 6
b) 7
c) 7.4
d) 8.2

Q.3. Poroscopy is ________.
a) counting pores of sweat glands only
b) counting pores of both sweat and sebaceous glands
c) counting number of ridges
d) counting number of ridges and sweat glands

Q.4. With tattoo marks destroyed, their presence can be inferred from the presence of pigment in the ________.
a) deep dermis
b) subcutaneous tissue
c) lymph nodes regional
d) underlying muscle

Q.5. A bluish discolouration of the neck of the tooth is due to ________.
a) cyanosis
b) bismuth

c) copper
d) nicotine

Q.6. For a suspected air embolism, the body cavity to be opened first is ________.
a) brain
b) thorax
c) abdomen
d) pelvis

Q.7. The 'under taker's fracture' due to falling on the head (backward) occurs at ________.
a) L5–S1
b) T12–L1
c) C6–C7
d) C1–C2

Q.8. The most fixed part of the intestine is ________.
a) duodenum
b) jejunum
c) colon
d) ileum

Q.9. After death, the blood usually remains fluid *except* in ________.
a) pneumonia
b) septicemia
c) CO poisoning
d) hypofibrinogenemia

Q.10. 'Nutmeg liver' refers to ________.
a) amoebic hepatitis
b) pyogenic abscess
c) chronic venous congestion
d) portal cirrhosis

Q.11. A child's brain attains a mature size and weight at about ________.
a) 3 yrs
b) 5 yrs
c) 7 yrs
d) 9 yrs

Q.12. Radiological signs of fetal death includes all *except* ________.
a) overlapping of skull bones (Spalding's sign)
b) hyperextension of the spine
c) collapse of the spinal column
d) gas in the aorta

Q.13. For transplantation, the cornea can be removed from the dead up to ________.
a) 6 hrs
b) 12 hrs
c) 18 hrs
d) 24 hrs

Q.14. Exhumation is done under order by ________.
a) police officer
b) superintendent police
c) first-class magistrate

Q.15. Diagnosis of brain death is dependent upon all *except* ________.
a) dilated/fixed pupil
b) no spontaneous breathing
c) flat EEG
d) cessation of cardiac contraction

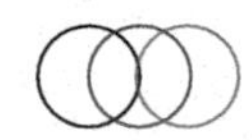

Q.16. After stoppage of circulation, muscles can live up to ________.
a) 10 min
b) 30 min
c) 3 hrs
d) 6 hrs

Q.17. Anoxic anoxia is produced by all *except* ________.
a) drowning
b) fire accidents
c) cyanide poisoning
d) strangulation

Q.18. Tardieu spots in hanging are common at all the following sites *except* ________.
a) scalp
b) eyebrow
c) chest wall
d) face

Q.19. Tardieu's spot is seen in which of the following conditions.
a) Septicemia
b) Endocarditis
c) Meningococcemia
d) All of the above

Q.20. The earliest sign of death is ________.
a) loss of skin elasticity
b) corneal clouding
c) cooling of body
d) postmortem lividity

Q.21. Tache noire refers to ________.
a) postmortem staining
b) flaccidity of eyeball
c) wrinkled, dusty sclera
d) maggot growth

Q.22. Rectal temperature does *not* appreciably fall till what time after death?
a) 15–30 min
b) 30–60 min
c) 60–90 min
d) None of the above

Q.23. Postmortem caloricity is seen in poisoning from ________.
a) arsenic
b) strychnine
c) cyanide
d) organophosphorus

Q.24. Postmortem lividity is well developed within ________.
a) 2 hrs
b) 4 hrs
c) 6 hrs
d) 8 hrs

Q.25. Fixation of postmortem staining occurs in ________.
a) 2 hrs
b) 4 hrs
c) 6 hrs
d) 8 hrs

Q.26. Rigor mortis starts when muscle ATP is reduced below ________.
a) 50%
b) 5%
c) 15%
d) 5%

Q.27. Rigor mortis is first evident in the ________.
a) intestine
b) myocardium
c) interstitial muscle
d) eyelids

Q.28. Cutis anserina of rigor mortis is due to stiffness of ________.
a) arrectores pilorum
b) biceps
c) cremaster muscle
d) diaphragm

Q.29. Rigor mortis does *not* occur in a fetus less than ________.
a) 9 mos
b) 7 mos
c) 6 mos
d) 8 mos

Q.30. The commonest cause of impotence in males is ________.
a) adrenal dysfunction
b) testicular failure
c) mal developed penis
d) psychogenic

Q.31. A cadaveric spasm is commonly seen in ________.
a) legs
b) hands
c) neck muscles
d) involuntary muscles

Q.32. Heat stiffening occurs when the body is exposed to temperature:
a) 450°C
b) 550°C
c) 650°C
d) 750°C

Q.33. The chief agent for bacterial putrefaction is ________.
a) *E. coli*
b) *B. fragilis*
c) *C. welchii*
d) *Staph aureus*

Q.34. Postmortem haemolysis is due to the bacterial enzyme
a) Lecithinase
b) Phospholipase
c) Streptokinase
d) Hyaluronidase

Q.35. The first external sign of putrefaction of a body lying in air is around the ________.
a) umbilicus
b) Rt iliac fossa
c) Lt iliac fossa
d) chest wall

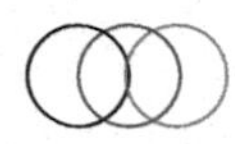

Q.36. A greenish colour is the earliest sign of putrefaction due to ________.
a) Hb
b) MetHb
c) SulfamethHb
d) CarboxyHb

Q.37. Combustible gas of autolysis is ________.
a) nitrogen dioxide
b) hydrogen sulfide
c) methane
d) carbon dioxide

Q.38. Postmortem luminescence is due to ________.
a) *Photobacterium fischeri*
b) *Armillaria mellea*

Q.39. The first internal organ to putrefy is the ________.
a) heart
b) brain
c) larynx/trachea
d) kidney

Q.40. The last organ to putrefy is the ________.
a) uterus/prostate
b) testes
c) ovary
d) adrenals

Q.41. Putrefaction occurs more rapidly in ________.
a) water
b) air
c) soil
d) cold saltwater

Q.42. Adipocere starts early in all of the following *except* ________.
a) face
b) breast
c) buttock
d) chest wall

Q.43. Dehydration/shriveling of a cadaver is called ________.
a) putrefaction
b) mummification
c) saponification

Q.44. Formaldehyde injected for embalming is ________.
a) 10%
b) 20%
c) 30%
d) 40%

Q.45. Odour of a mummified body is ________.
a) pungent
b) putrid
c) offensive
d) odourless

Q.46. For embalming, chemicals are injected into the ________.
a) femoral artery
b) abdominal cavity
c) chest cavity

Q.47. Maggots appear in natural orifices of dead bodies in the summer in about ________.
a) 2–4 hrs
b) 6–8 hrs
c) 8–12 hrs
d) 12–24 hrs

Q.48. After death, all of following show a rise in CSF *except* ________.
a) lactic acid
b) amino acid
c) urea
d) uric acid

Q.49. After death, the blood level of the following decreases:
a) Sodium
b) Potassium
c) Magnesium
d) None

Q.50. A brush burn refers to ________.
a) electric burn
b) lightning burn
c) sliding abrasion
d) pressure abrasion

Answers (Exercise 4)

1. b	2. c	3. a	4. c	5. d	6. b	7. c	8. a	9. a	10. c
11. c	12. b	13. d	14. a	15. b	16. d	17. c	18. c	19. d	20. a
21. c	22. c	23. b	24. b	25. c	26. c	27. d	28. a	29. b	30. d
31. b	32. c	33. c	34. a	35. b	36. c	37. b	38. a	39. c	40. a
41. c	42. d	43. b	44. d	45. d	46. a	47. d	48. c	49. a	50. c

Exercise 5

Q.1. A wound caused by a sickle is ________.
a) stab wound
b) incised wound
c) both

Q.2. A bevelling cut refers to ________.
a) flap wound
b) overhang margins
c) one margin undermined
d) irregular margins

Q.3. A cutthroat wound causes death by ________.
a) air embolism
b) haemorrhage
c) tracheobronchial aspiration
d) all of the above

Q.4. Blunt trauma more likely to produce incised-like wounds in the ________.
a) hand
b) neck
c) chest
d) scalp

Q.5. Fabricated wounds are mostly ________.
a) abrasions
b) incised wounds
c) contusions
d) lacerations

Q.6. A diastatic fracture refers to a fracture through ________.
a) outer table
b) inner table
c) sutural line
d) none of the above

Q.7. In a countercoup injury, impact or injury is observed ________.
a) at site of impact
b) at a site opposite to impact
c) at a site tangential to impact

Q.8. An extradural haemorrhage commonly occurs from the rupture of the ________.
a) superior sagittal sinus
b) middle meningeal artery
c) dural sinus
d) none of the above

Q.9. The common cause of death in extradural haemorrhage is ________.
a) hemorrhagic shock
b) respiratory failure
c) cardiac failure
d) neurogenic shock

Q.10. Rupture of berry aneurysm commonly produces ________.
a) subarachnoid bleed
b) intracerebral sued
c) subdural bleed

Q.11. The most common cause of aneurysm formation is ________.
a) trauma
b) septic emboli
c) congenital
d) syphilitic endarteritis

Q.12. Whiplash is which form of injury?
a) Hyperflexion
b) Hyperextension
c) Lateral flexion
d) Atlantoaxial dislocation

Q.13. In blunt abdominal trauma, the commonest site of GI ruptures is the ________.
a) stomach
b) duodenum
c) jejunum
d) transverse colon

Q.14. In drowning, the epidermis of the hands and feet is separated in the form of gloves and stocking after ________.
a) 2 minutes
b) 2 hours
c) 2 weeks
d) 2 months

Q.15. Chadwick's sign is ________.
a) softening of cervix
b) increased vaginal mucous secretion
c) blue colouration of vagina

Q.16. Amount of air necessary to produce fatal air embolism is ________.
a) 20 mL
b) 50 mL
c) 100 mL
d) 250 mL

Q.17. Malignant hyperthermia is a danger with ________.
a) atropine
b) succinylcholine
c) pancuronium

Q.18. Hypothermia is said to exist when rectal/oral temp is less than ________.
a) 35°C
b) 30°C
c) 25°C
d) 20°C

Q.19. Frostbite occurs when there is a continuous exposure to temperature in the range of ________.
a) –10°C and below
b) –2.5°C and below
c) 2–4°C
d) 5–10°C

Q.20. Frostbite is very common in ________.
a) lips
b) nose
c) cheeks
d) hair

Q.21. The burn type that is relatively painless is ________.
a) 1st degree
b) 2nd degree
c) 3rd degree

Q.22. Minimum temp to produce a burn is ________.
a) 40°C
b) 44°C
c) 50°C
d) 60°C

Q.23. Hemoglobinuria occurs when the burnt skin surface exceeds ________.
a) 20%
b) 30%
c) 50%
d) 70%

Q.24. Scalding is caused when liquid in contact with skin has a temperature above ________.
a) 44°C
b) 50°C
c) 60°C
d) 80°C

Q.25. Satyriasis is ________.
a) excessive inclination for sex
b) dislike for sex
c) sexual perversion

 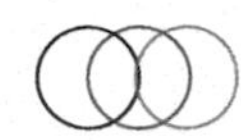

Q.26. The most reliable method for personal identification is ________.
a) photography
b) handwriting
c) anthropometry
d) dactylography

Q.27. The age of full criminal responsibility is ________.
a) 7 years
b) 18 years
c) 21 years
d) 30 years

Q.28. A person died from a stab wound in the aorta in winter. His rectal temperature was 31°C. The postmortem interval is ________.
a) 2 hours
b) 3 hours
c) 4 hours
d) 8 hours

Q.29. The presence of tache noire is suggestive that the time since death is ________.
a) 1 hour
b) 2 hours
c) 3 hours
d) 8 hours

Q.30. The last organ to putrefy in young girls is ________.
a) uterus
b) bladder
c) ovary
d) heart

Q.31. Exhumation is ________.
a) artificial preservation of a dead body
b) burning of a dead body
c) getting the dead body from the grave
d) aseptic autolysis of a dead body

Q.32. Flattening remains until ________.
a) hypostasis is complete
b) rigor mortis is complete
c) putrefaction begins
d) cadaveric spasm occurs

Q.33. A person died in a hospital 2 weeks after the burning of his abdominal and chest walls and upper extremities. The most probable cause of death is ________.
a) traumatic asphyxia
b) pulmonary fat embolism
c) perforation of an acute duodenal ulcer (curling's ulcer)
d) suprarenal haemorrhage

Q.34. Polar fracture is a ________.
a) comminuted fracture
b) fissure fracture
c) depressed fracture
d) cut fracture

Q.35. The most resistant body tissue to electrical injury is ________.
a) dry skin
b) muscles
c) bones
d) blood and body fluid

Q.36. The commonest cause of death in extradural haemorrhage is ________.
a) hemorrhagic shock
b) respiratory failure
c) cardiac failure
d) neurogenic shock

Q.37. A 32-year-old farmer received abdominal trauma; after 2 hours, the pulse reached 136/min, blood pressure 70/40. The cause is ________.
a) shock
b) sympathetic shock
c) internal haemorrhage
d) air embolism

Q.38. The most important sign for identification of contact firearm inlet is ________.
a) loss of substance
b) presence of two wounds
c) muzzle imprint around the wound margin
d) abraded inlet

Q.39. A person with a head injury can talk normally and tell about the circumstantial evidence in case of ________.
a) lucid interval
b) concussion
c) retrograde amnesia
d) automatism

Q.40. A tear in the intima of the carotid artery with bleeding into its wall is seen in cases of ________.
a) smothering
b) ante mortem hanging
c) postmortem hanging
d) traumatic asphyxia

Q.41. One of the following is a sure external sign of drowning that can be found in postmortem examination:
a) Goose skin
b) Washer woman's hands
c) Peeling of the skin
d) Fine froth at the mouth and nostrils

Q.42. One of the following is *not* the immediate cause of death in criminal abortion:
a) Haemorrhage
b) Septic instrumentation
c) Reflex vagal inhibition of the heart
d) Air embolism

Q.43. Choose the *wrong* statement regarding child abuse:
a) Injuries arc deliberate.
b) There is an inconsistent history for injuries.
c) Stepparents are rarely involved.
d) There is a slight tendency toward the male child.

Q.44. Adipocere of the whole body occurs in about ———.
a) 1 month
b) 2 months
c) 6 months
d) 12 months

Q.45. Dry burn is caused by ________.
a) hot liquid or steam
b) flame or hot metals

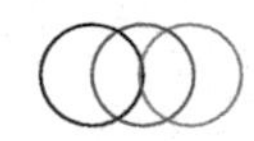

- c) deep X-ray or UV-rays
- d) strong acids or alkalis

Q.46. One of the following is *not* a manifestation of shaken baby syndrome:
- a) Burns
- b) Retinal haemorrhage
- c) Rib fracture
- d) Subdural hematoma

Q.47. One of the followings produce toxic hypothermia:
- a) Salicylates
- b) Anticholinergics
- c) Antidepressants
- d) Opioids

Q.48. The best method to avoid aspiration of fluids during gastric lavage in a comatose patient is by ________.
- a) putting the head of the patient at a lower level than his feet
- b) putting the patient in the left lateral position
- c) introduction of a cuffed endotracheal tube before lavage
- d) continuous suction of the fluid from the trachea

Q.49. Surgical interference may be needed especially in children after poisoning by ________.
- a) hydrocyanic acid
- b) caustic potash
- c) hydrochloric acid
- d) carbolic acid

Q.50. One of the following solvents is *not* metabolized in the body to cyanide:
- a) Isopropanol
- b) Nitroprusside
- c) Acetonitrile
- d) Acrylonitrile

Q.51. In poisoning with hydrocyanic acid, nitrates are given in order to ________.
- a) reduce cyanide
- b) induce vasodilatation
- c) produce methemoglobin
- d) oxidize cyanide

Q.52. After skin contamination, the patient passed into a coma with miosis and finally acute nephritis. The poison is ________.
- a) oxalic acid
- b) nitric acid
- c) hydrocyanic acid
- d) carbolic acid

Q.53. The dose of sodium thiosulfate for treatment of cyanide poisoning in children is ________.
- a) 112.5 mg/kg IV over 10–20 min
- b) 412.5 mg/kg IV over 10–20 min
- c) 412.5 mg/kg IV over 2 min
- d) 412.5 mg/kg IV over 5 min

Q.54. An old traffic policeman in a busy street of Cairo is liable to suffer from ________.
- a) spastic gait
- b) tremors
- c) masked face
- d) wrist and ankle drop

Q.55. A blue line in the gingival margin in the case of lead poisoning is due to the deposition of ________.
- a) lead chromate
- b) lead sulfide
- c) lead subacetate
- d) lead iodide

Q.56. In iron poisoning, bloody vomiting and diarrhoea, massive fluid loss in GIT, renal failure, and death occur in ________.
- a) Stage 1
- b) Stage 2
- c) Stage 3
- d) Stage 4

Q.57. The specific antidote in the case of iron poisoning is ________.
- a) DMSA
- b) Desferrioxamine
- c) EDTA
- d) Penicillamine

Q.58. Acute toxicity of organophosphates causes ________.
- a) urine retention
- b) oliguria
- c) urine incontinence
- d) anuria

Q.59. Prolonged prothrombin time occurs in cases of poisoning with ________.
- a) Parathion
- b) Warfarin
- c) Paraquat
- d) zinc sulfide

Q.60. The second stage of acute acetaminophen toxicity is characterized by ________.
- a) abnormalities of liver function tests
- b) bleeding tendencies due to coagulation defect
- c) nausea and malaise
- d) right-upper-quadrant pain and tenderness

Q.61. Which of the following is a specific antidote for acute acetaminophen toxicity?
- a) BAL
- b) Mucomyst
- c) EDTA
- d) DMSA

Q.62. Benzodiazepines act on the CNS through the following mechanism:
- a) Increasing catecholamines
- b) Increasing serotonin
- c) Increasing the activity of GABA
- d) Decreasing the activity of GABA

Q.63. In the case of foodborne botulism, the toxin is ________.
- a) formed in the duodenum
- b) formed in the colon
- c) formed in the intestine
- d) formed in the canned food before consumption

Q.64. A four-year-old boy accidentally ingested a clear fluid, vomited twice, and then started to cough with tachypnea. 24 hours later he developed a fever of 39°C due to bronchopneumonia. The possible diagnosis is ________.
- a) phenol toxicity
- b) kerosene toxicity

 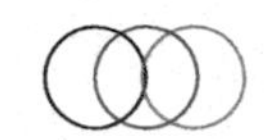

c) ethanol toxicity
d) methanol toxicity

Q.65. Macewen's sign is a manifestation of a massive intake of ________.
a) atropine
b) opium
c) methanol
d) ethanol

Q.66. One of the following manifestations is an indication of severe ethanol intoxication:
a) Euphoria and sense of well-being
b) Marked muscular incoordination
c) Increased confidence
d) Aggressive behaviour

Q.67. In the case of acute CO poisoning, coma and death with lively red colour occur at a carboxy haemoglobin level of ________.
a) 10%–20%
b) 20%–30%
c) 30%–40%
d) 50%–60%

Q.68. Pathological jealousy is diagnostic of ________.
a) cocaine intoxication
b) cannabis intoxication
c) alcoholic intoxication
d) tobacco intoxication

Answers (Exercise 5)

1. c	2. c	3. d	4. d	5. b	6. c	7. b	8. b	9. b	10. a
11. d	12. b	13. d	14. d	15. c	16. c	17. b	18. a	19. b	20. b
21. c	22. b	23. b	24. c	25. a	26. a	27. b	28. c	29. c	30. a
31. c	32. c	33. c	34. b	35. c	36. b	37. c	38. c	39. a	40. b
41. d	42. b	43. d	44. c	45. b	46. a	47. d	48. c	49. b	50. a
51. c	52. d	53. b	54. d	55. b	56. a	57. b	58. c	59. b	60. a
61. b	62. c	63. d	64. b	65. d	66. b	67. d	68. c		

17.2 TRUE OR FALSE STATEMENTS

(Write T for True or F for False.)

Exercise 6

Q.1. Porphyria is a congenital type of photosensitization.
Q.2. Spraying of 2,4-D in sugar beets is a predisposing factor for cyanide poisoning.
Q.3. Kidney is the tissue of choice for detection of most metals and sulfonamides.
Q.4. ANTU is less toxic when stomach is full than when it is empty.
Q.5. The toxicity of fluorocitrate is due to its conversion in body to highly toxic fluoroacetate.
Q.6. Biphasic reaction is a characteristic of carbamate poisoning.
Q.7. Cyanide poisoning could be detected if fresh rumen content contains 10 ppm HCN.
Q.8. Ochratoxin is primarily a hepatotoxic mycotoxin.
Q.9. Vitamin K is the best treatment for an animal showing signs of shock from Warfarin toxin.
Q.10. Environmental toxicology does not deal with toxic effects on humans.
Q.11. Intravenous route is the fastest route of exposure for direct-acting drugs.
Q.12. Methamphetamine doses (4 × at 5 mg/kg) always result in death.
Q.13. Potentiation is when two agents that have an effect increase the effect when combined.
Q.14. Chemical antagonism is also referred to as inactivation.
Q.15. Ipecac syrup and activated charcoal combination is most effective.
Q.16. Warm water with mustard can be used to induce emesis.
Q.17. Gastric lavage can be performed in some unconscious patients.

Answers (Exercise 6)

1. T 2. F 3. T 4. F 5. F 6. F 7. F 8. F 9. F 10. F
11. T 12. F 13. F 14. T 15. F 16. T 17. T

17.3 FILL IN THE BLANKS

Exercise 7

Q.1. The type of evidence seen at the time of poisoning is referred as ________.

Q.2. The most common feed contaminant that can be expected during improper storage is ________.

Q.3. Pink colouration in urine is suggestive of poisoning due to ________.

Q.4. Phenols and cresols produce ________ colouration of urine.

Q.5. The symptoms or lesions that is characteristic to a particular toxicant is known as ________.

Q.6. The evidence that is obtained during postmortem examination is known as ________ evidence.

Q.7. Bitter almond smell of ruminal contents is suggestive of ________.

Q.8. Poisoning with phosphorus results in ________ odour during postmortem examination.

Q.9. The detection of toxic material in the body using laboratory methods constitutes ________ evidence.

Q.10. The evidence that is obtained by feeding suspected material (feed) to healthy animals to ascertain the presence of a toxicant is ________.

Q.11. The aim of treatment during poisoning is to ________ the threshold of the toxicant.

Q.12. The time versus concentration curve of a toxicant in the body is shaped ________.

Q.13. The ascending phase of the time-concentration of a toxicant curve represents ________.

Q.14. The descending phase of the time-concentration of a toxicant curve represents ________.

Q.15. When emesis is contraindicated, the safest alternative is ________.

Q.16. The most commonly used adsorbing agent to bind toxicants in the GIT is ________.

Q.17. The type of diuretics or purgatives that are preferred in cases of poisoning is ________.

Q.18. When a large amount of toxicant is absorbed into the body or when renal failure occurs ensues, the method of choice employed for elimination of the toxicant is ________.

Q.19. The mechanism involved in the enhanced elimination of acidic agents in alkalized urine and basic agents in acidified urine is ________.

Q.20. The substance that counteracts or neutralizes a toxicant is known as ________.

Answers (Exercise 7)

1. Circumstantial evidence
2. Mycotoxins (Aflatoxin)
3. Phenothiazines
4. Green
5. Pathognomonic (symptom or lesion)
6. Pathological
7. Cyanide poisoning
8. Garlic-like
9. Analytical
10. Experimental evidence.
11. Increase
12. Bell or inverted 'U'
13. Absorption
14. Excretion
15. Gastric lavage
16. Activated charcoal
17. Osmotic/saline type

18. Dialysis
19. Ion trapping
20. Antidote

Exercise 8

Q.1. The most common route of accidental exposure to toxicants in toddlers is ________.
Q.2. Gray baby syndrome is due in part to the toxic response to ________ and ________.
Q.3. The purposely fatal intoxication of an individual is investigated by ________.
Q.4. A toxicant that comes into contact with an individual through the skin can be metabolized through the ________.
Q.5. The toxic effect of chronic exposure to benzene is ________ and acute ________.
Q.6. The increased toxicity of normal young children to certain toxicants is due to ________ not fully developed and lack of ________.
Q.7. The combination of carbon tetrachloride with ethanol on liver toxicity results in ________.
Q.8. Two compounds that present an effect of 2 + 2 = 6 indicates that both are ________ and have ________ effect.
Q.9. When two substances act on the same physiological function with opposing effects, the effect will be ________.
Q.10. The treatment of an overdose of anticoagulant Warfarin with phenobarbital is an example of ________.
Q.11. An example of a therapy involving receptor antagonism is ________.
Q.12. Two compounds that present an effect of 4 + (–4) = 0 indicates that ________.
Q.13. When two chemicals antagonize, it means ________.
Q.14. Ethanol acts as an antidote for methanol poisoning by ________.
Q.15. The antidote for alkaloids is ________.
Q.16. The antidote for paracetamol (acetaminophen) toxicity is ________.
Q.17. The agents that are used in the case of CNS depression and respiratory arrest are called ________.
Q.18. Secondary photosensitization in *Lantana camara* is mainly due to ________.
Q.19. The phytoconstituents of *L. camara* that cause bile-duct occlusion and liver damage are ________.
Q.20. Hepatic lesions such as ________ and ________ are diagnostic in differentiation primary and secondary photosensitization.
Q.21. Plasma from blood can be prepared by adding ________ in the blood.
Q.22. Serum can be prepared by letting the collected patient's blood clot. The resulting supernatant is called ________.
Q.23. Elapidaes (snakes) are ________.

Answers (Exercise 8)

1. Ingestion
2. Chloramphenicol, immature kidneys
3. Forensic toxicologist
4. Liver
5. Leukaemia, CNS depression
6. Blood–brain barrier, metabolizing enzymes
7. Synergism [more than additive (2 + 2=10)]
8. Toxic, synergistic effect
9. Antagonism
10. Dispositional antagonism
11. The use of tamoxifen in cancer
12. Antagonism

13. Opposite effect
14. Competitive inhibition (mechanism)
15. Potassium permanganate
16. *N*-acetyl cysteine
17. Analeptics
18. Bile duct occlusion
19. Lantadene A & B
20. Fibrosis and biliary hyperplasia
21. Anticoagulant
22. Serum
23. Neurotoxic

17.4 MATCH THE STATEMENTS

(Column A with Column B)

Exercise 9

Column A			Column B
Q.1.	basophilic stippling	a.	BAL
Q.2.	penicillamine chelates	b.	atropine
Q.3.	drug of choice in carbamate poisoning	c.	copper
Q.4.	rubratoxin	d.	estrogenic mycotoxin
Q.5.	antidote in nitrate poisoning	e.	lameness
Q.6.	arsenic	f.	lead
Q.7.	zearalenone	g.	hepatic cirrhosis
Q.8.	sorghum species	h.	cyanogenic glycoside
Q.9.	pyrrolizidine alkaloid	i.	*Penicillium rubrum*
Q.10.	fluorosis	j.	methylene blue

Answers (Exercise 9)

1. f 2. c 3. b 4. i 5. j 6. a 7. d 8. h 9. g 10. e

Exercise 10

Column A (poison)		Column B (antidote/antagonist)	
Q.1.	arsenic	a.	atropine/pralidoxime
Q.2.	cyanide	b.	Fuller's
Q.3.	methanol	c.	dimercaprol
Q.4.	paraquat	d.	ethanol
Q.5.	parathion	e.	sodium nitrite/sodium thiosulfate
Q.6.	lead	f.	*N*-acetyl cysteine
Q.7.	paracetamol	g.	EDTA
Q.8.	LSD	h.	gen-metal kidney
Q.9.	thalidomide	i.	abuse
Q.10.	copper	j.	teratogenicity

Answers (Exercise 10)

1. c 2. e 3. d 4. b 5. a 6. g 7. f 8. i 9. j 10. h

Section 9
SPECIAL TOPICS

CHAPTER 18

ADVERSE EFFECTS OF CALORIES

18.1 MULTIPLE-CHOICE QUESTIONS

(Choose the most appropriate response.)

Exercise 1

Q.1. Which of the following definitions is *false*?

a) The set-point hypothesis proposes that food intake and energy expenditure are coordinately regulated by defined regions in the brain that signal to maintain a relatively constant level of energy reserve and body weight.
b) Hormonal messages generated by the endocrine cells of the pancreas, adipose tissue, and GI tract are involved in orchestrating multiple responses associated with caloric intake and caloric utilization.
c) Caloric content of foods generally assumes factors of 4, 9, and 4 for carbohydrate, fat, and protein, respectively.
d) The body mass index (BMI) is an accurate method for assessing body composition.
e) Liver, adipose, muscle, and other tissues adapt to excess caloric loads.

Q.2. Neural control of energy balance ________.

a) may be defined as the action of leptin on CNS function
b) may be defined as the action of hypothalamic cholinergic control of appetite and hedonic control
c) may involve a balance between food intake and energy expenditure
d) may involve a balance between leptin's action on orexigenic versus anorexigenic peptide expression
e) may involve adrenocortical control of hepatic unction

Q.3. Body composition may be assessed by ________.

a) electrical impedance because lean mass has more water and greater conductivity than at mass
b) anthropometric analysis of the body mass index
c) hydrodensitometry, which uses the density for the whole body and corrects or residual air in the lungs and GI tract to determine relative body fat
d) nuclear magnetic resonance
e) all of the above

Q.4. Ectopic at deposition includes ________.

a) adipose tissue
b) skeletal muscle
c) lungs
d) heart
e) GI tract

Q.5. Excess calories may be ________.

a) stored as glucose in adipose tissue
b) stored as triglycerides in CNS tissue
c) stored as glycogen in CNS tissue
d) stored as glycogen in the liver
e) stored as triglycerides in the GI tract

Q.6. Metabolic syndrome is a constellation of actions including ________.

a) typical results from elevated fasting glucose, increased HDL, and hypertension
b) typical results from elevated fasting glucose, increased LDL, and hypertension
c) typical results from elevated fasting glucose, hypertriglyceridaemia, and hypotension
d) typical results from elevated fasting glucose, hypotriglyceridaemia, and truncal obesity
e) typical results from elevated fasting glucose, hypertriglyceridaemia, and truncal obesity

Q.7. Excess caloric intake ________.

a) may lead to non-alcoholic steatohepatitis
b) is always correlated with obesity and insulin resistance
c) is characterized by elevations of serum AL concentrations in all cases
d) leads to hepatic cirrhosis and liver cancer in almost all cases
e) is readily reversible by dieting

Q.8. Although dieting may effectively reduce body weight, ________.

a) toxicity may result from stimulation of adipokine release.
b) toxicity may result from inhibition of drug-metabolizing enzymes.
c) toxicity may result from a loss of required nutrients.
d) toxicity may result from extreme mental illness.
e) toxicity may result from weight cycling.

Q.9. Body mass index ________.

a) may be used as an indicator of sufficient caloric and essential nutrient intake
b) may be defined as body height divided by body weight squared
c) has risen insignificantly over the past 30 years in the United States
d) may not be used in the estimation of cancer risk in humans
e) may be defined as body weight divided by height squared

Q.10. Humans consume food to provide energy needed to ________.

a) drive cellular functions including digestion, metabolism, pumping blood, nerve activity, and muscle contractions
b) promote photosynthesis
c) synthesize oxygen in the lungs
d) prepare minerals for use in the body
e) produce carbon dioxide to fuel body functions

Answers (Exercise 1)

1. d 2. d 3. e 4. b 5. d 6. e 7. a 8. c 9. e 10. a

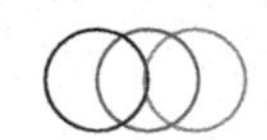

18.2 TRUE OR FALSE STATEMENTS

(Write T for True or F for False.)

Exercise 2

Q.1. Energy in the body is derived from one main nutrient made up of sugar.
Q.2. Hormonal messages generated by the pancreas, adipose tissue, and GI tract orchestrate multiple responses associated with caloric intake and utilization.
Q.3. Minerals found in food are the main ingredients that are necessary for proper growth, development, reproduction, and repair.
Q.4. The "set-point" hypothesis proposes that food intake and energy expenditure are coordinately regulated by liver to maintain a relatively constant level of energy reserve and body weight.
Q.5. All biotic and non-biotic organisms derive energy from food to sustain life.
Q.6. Brain has little to no stored energy in the form of glycogen or triglycerides.
Q.7. Accurate assessment of the caloric value of foods is essential for effective nutritional management in clinical and public policy arenas.
Q.8. Lean mass includes protein, carbohydrate, and minerals.
Q.9. Positive energy balance produced as a result of overeating and inadequate physical activity result in toxicity over the long term.
Q.10. Triglycerides and glycogen are used by the body to store excess caloric energy.

Answers (Exercise 2)

1. F 2. T 3. F 4. F 5. F 6. T 7. T 8. T 9. T 10. T

CHAPTER 19

TOXIC EFFECTS OF NANOPARTICLES

19.1 MULTIPLE-CHOICE QUESTIONS

(Choose the most appropriate response.)

Exercise 1

Q.1. The goals of nanotoxicology are ________.
- a) to identify and characterize hazards of engineered nanomaterials
- b) to determine "safe" exposure levels
- c) to determine biologic and biochemical actions
- d) to determine manufacturing procedures and cost
- e) to determine preventive exposure guidelines

Q.2. Which of the following answers is *not* true regarding nanoparticles (NPs)?
- a) NPs can originate from natural sources including forest fires, volcanoes, and viruses.
- b) NPs can originate from unintentional sources including internal combustion engines and electric motors.
- c) NPs can originate from unintentional sources including ferritin and magnetotactic bacteria.
- d) NPs can originate from intentional sources including carbon nanotubes and metal oxide nanoparticles.
- e) NPs can originate from natural, intentional, and unintentional anthropogenic sources.

Q.3. In contrast to larger particles > 500 nm, nanoparticles ________.
- a) are highly likely to enter the body by dermal absorption
- b) are highly likely to enter the body through the respiratory tract
- c) are unlikely to adsorb to protein or lipid
- d) are efficiently removed from the lungs via mucociliary transport
- e) are not likely to undergo uptake and transport in sensory neurons

Q.4. Which of the following statements is *not* true?
- a) Nanomaterials may be classified by geometry and chemistry.
- b) Engineered nanomaterials include quantum dots, C-nanofibre arrays, and few-layer grapheme.
- c) Agglomerates include primary particles held together by weak van der Waals forces.
- d) Aggregates include primary particles held by strong chemical bonds.
- e) Hydrodynamic diameter is unimportant in particle interactions.

Q.5. Nanoparticles can exert toxicity by all of the following mechanisms *except* ________.
- a) damage to DNA and chromosomes
- b) induction of oxidant stress
- c) interference with biotransformation enzyme activities
- d) activation of signaling pathways
- e) release of toxic metal ions from internalized NPs

Q.6. Biodistribution of nanoparticles may be influenced by ________.
- a) physicochemical properties such as plasma protein and respiratory tract mucus
- b) physicochemical properties such as surface size and chemistry
- c) physicochemical properties such as the gastrointestinal milieu
- d) body compartment media including surface hydrophobicity
- e) body compartment media including size

Q.7. Assays to determine the toxicity of manufactured nanoparticles suffer from all of the complications below *except* ________.
   a) the nanomaterial aggregate may no longer be in the nanosize range
   b) aggregates of the nanoparticle may settle out of solution, which may affect exposure dose
   c) alterations in surface chemistry to stabilize suspension may evoke other issues in toxicity assessment
   d) coatings of particles may have their own toxicity
   e) uptake of the nanoparticle into an organism is easily determined

Q.8. Which of the following is *not* a nanoparticle?
   a) Carbon nanotubes
   b) Bucky-ball
   c) Graphene
   d) Zinc nanorods
   e) Bacteria

Answers (Exercise 1)

1. d 2. c 3. b 4. e 5. c 6. b 7. e 8. e

 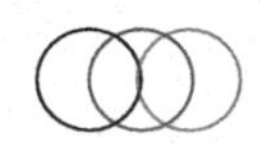

19.2 TRUE OR FALSE STATEMENTS

(Write T for True or F for False.)

Exercise 2

Q.1. Nanotoxicology is the study of adverse effects of nanomaterials on the environment only.

Q.2. Nanotechnology means nanoscale dimensions between approximately 100 and 1000 nm, where unique phenomena enable novel applications.

Q.3. Surface properties are major determinants of biologic reactivity due to high surface area, surface charge, dissolution and release of metal ions, and redox activity leading to generation of ROS.

Q.4. The respiratory tract is the major route for humans to exposure of nanomaterials.

Q.5. In nanotoxicology, surface properties are major determinants of biologic reactivity due to high surface area.

Q.6. Dosimetry means dose and route of administration.

Q.7. Nanomaterials concern particle size only and do not involve composition, geometry, and complexity of molecules.

Q.8. Nanomaterial structures, at the nanoscale, have high surface-to-volume or surface-to-mass ratios.

Answers (Exercise 2)

1. F 2. F 3. T 4. T 5. T 6. F 7. F 8. T

CHAPTER 20

OCCUPATIONAL TOXICOLOGY

20.1 MULTIPLE-CHOICE QUESTIONS

(Choose the most appropriate response.)

Exercise 1

Q.1. Which of the following infectious agents can cause hepatocellular carcinoma?
- a) Flavivirus
- b) Bunyavirus
- c) Alphavirus
- d) Hepatitis C virus
- e) Hepatitis B virus

Q.2. Which of the following is *least* likely to increase occupational inhalation of a chemical?
- a) Increased airborne concentration
- b) Increased respiratory rate
- c) Increased tidal volume
- d) Increased particle size
- e) Increased length of exposure

Q.3. Which would increase the likelihood of toxic dosage through dermal exposure?
- a) No pre-existing skin disease
- b) Toxic exposure to thick skin
- c) Increased percutaneous absorption rate
- d) Low surface area of exposure
- e) High epidermal intercellular junction integrity

Q.4. Prolonged arsenic exposure could cause ________.
- a) infertility
- b) cirrhosis
- c) cor pulmonale
- d) skin cancer
- e) nephropathy

Q.5. Which of the following lung diseases has the highest occupational death rate?
- a) Asbestosis
- b) Coal workers' pneumoconiosis
- c) Byssinosis
- d) Hypersensitivity pneumonitis
- e) Silicosis

Q.6. Lyme disease is caused by which of the following?
- a) *B. burgdorferi*
- b) *H. capsulatum*
- c) *M. tuberculosis*
- d) *L. pneumophila*
- e) *C. psittaci*

Q.7. Asbestos exposure is unlikely to cause ________.
   a) lung cancer
   b) GI cancer
   c) emphysema
   d) pulmonary fibrosis
   e) mesothelioma

Q.8. Exposure to which of the following can cause autoimmune disease?
   a) Mercury
   b) Nitrogen dioxide
   c) Vinyl chloride
   d) Lead
   e) Flavivirus

Q.9. Which of the following might be linked to parkinsonism?
   a) Nitrogen dioxide
   b) Zinc
   c) Copper
   d) Magnesium
   e) Carbon monoxide

Q.10. Which of the following is *not* a modifying actor that can influence the likelihood of disease?
   a) Age
   b) Dose
   c) Nutritional status
   d) Gender
   e) Genetic susceptibility

Q.11. What is the expected response of an individual exposed to a single absorbed dose of 10 rads (0.1 Gy) of whole-body X-irradiation?
   a) Severe bone marrow depression
   b) Permanent sterilization
   c) No adverse response
   d) Vomiting

Q.12. What form of mercury was the cause of Minamata Bay disease?
   a) Mercuric salts
   b) Mercurous salts
   c) Organic mercury
   d) Elemental mercury

Q.13. What was the regulatory response to the Delaney clause?
   a) Permitted most food additives to be declared generally recognized as safe (GRAS)
   b) Prohibited EPA from setting safe exposure levels for environmental carcinogens
   c) Prohibited FDA from approving food additives found to cause cancer in animals
   d) Was applied only to food additives demonstrating human evidence of carcinogenicity

Answers (Exercise 1)

1. e	6. a	11. c
2. d	7. c	12. c
3. c	8. c	13. c
4. b	9. e	
5. b	10. b	

 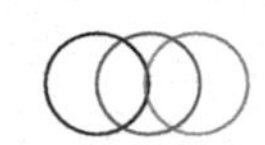

20.2 TRUE OR FALSE STATEMENTS

(Write T for True or F for False.)

Exercise 2

Q.1. In occupational environments, exposure is through inhalation only.
Q.2. Occupational toxicology is the risk assessment towards chemical and biologic hazards.
Q.3. In occupational toxicology, it is often difficult to establish a causal link between a worker's illness and job.
Q.4. For chemical and biological agents, exposure limits are expressed as acceptable ambient concentration levels.
Q.5. ADIs are used as established standards by regulatory agencies or as guidelines by occupational trade or research groups.
Q.6. OELs correspond to the level of exposure below which the probability of impairing the health of the exposed workers is acceptable.
Q.7. Acute pulmonary oedema, bronchiolitis obliterans is caused by nitrogen oxides.
Q.8. Hepatic hemangiosarcoma is caused by benzene.
Q.9. Occupational diseases of the reproductive system can be gender- and organ-specific, or may affect both sexes.
Q.10. Animal studies provide valuable data at which the risk of health impairment is acceptable.

Answers (Exercise 2)

1. F 2. F 3. T 4. T 5. F 6. T 7. T 8. F 9. T 10. T

CHAPTER 21

VETERINARY DRUG RESIDUE HAZARDS

21.1 MULTIPLE-CHOICE QUESTIONS

(Choose the most appropriate response.)

Exercise 1

Q.1. A concentration of 0.01% is equivalent to how many parts per million (ppm)?
   a) 1 ppm
   b) 10 ppm
   c) 100 ppm
   d) 1000 ppm
   e) 10,000 ppm

Q.2. Heterocyclic amines and acrylamide are food contaminants that ________.
   a) are produced by microorganisms
   b) are produced by the process of cooking
   c) are considered as GRAS
   d) are residues from animal feeds

Q.3. The concepts of "de minimis" as applied to food safety means
   a) Find the smallest harmful dose
   b) Only food colours 1/100 of the no-observed adverse-effect level (NOAEL) can be used
   c) Pesticide residues can be present at the acceptable daily intake (ADI)
   d) The risk is so small it is of no concern

Q.4. Which category of insecticidal compounds presents a problem of persistent residues in fatty tissues of animals?
   a) Carbamates
   b) Organochlorines
   c) Organophosphates
   d) Pyrethrins
   e) Juvenile hormones

Q.5. The most common type of drug residue in animal products is ________.
   a) antibiotic
   b) metal
   c) aflatoxin
   d) none

Answers (Exercise 1)

1. c 2. b 3. d 4. b 5. a

21.2 TRUE OR FALSE STATEMENTS

(Write T for True or F for False.)

Exercise 2

Q.1. The major public health significances of drug residue help in better growth.
Q.2. Safe concentration is arrived by dividing ADI with the amount of edible organ that is generally consumed in a day.
Q.3. The tolerance for the extra label use of a drug is zero.
Q.4. Withdrawal time can be extrapolated from the area under curve calculated from pharmacokinetic studies.
Q.5. Pesticides are the most frequently found drug residues followed by anti-inflammatory drugs.
Q.6. Drug residues do *not* cause any allergic reaction in sensitive individuals.

Answers (Exercise 2)

1. F 2. T 3. T 4. F 5. F 6. F

 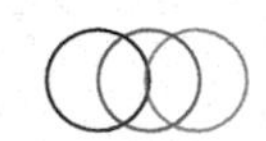

21.3 FILL IN THE BLANK STATEMENTS

Exercise 3

Q.1. Heterocyclic amines and acrylamide are food contaminants that are produced by the process of ________.

Q.2. Category of insecticidal ________ compounds present a problem of persistent residues in the fatty tissues of animals.

Q.3. The tolerance for extra label use of a drug is ________.

Q.4. The maximum acceptable/permitted amount of a drug present in feed and foods is known as ________.

Q.5. The highest dose of a compound that produces adverse effects but no mortality is called ________.

Answers (Exercise 3)

1. Cooking
2. Organochlorines
3. Zero
4. Maximum residue level (MRL) (for pesticides – maximum residue limit)
5. Maximum tolerated dose or minimum toxic dose (MTD)

FURTHER READING

Aiello, Susan E. (2016). *The Merck Veterinary Manual*, 11th ed. Merck & Co Inc, Kenilworth, NJ.

Beasley, V. (1999). Absorption, distribution, metabolism, and elimination: Differences among species. In: *Veterinary Toxicology*, Ed. V. Beasley, International Veterinary Information Service (www.ivis.org), Ithaca, NY. https://www.researchgate.net/publication/268347521_Absorption_Distribution_Metabolism_and_Elimination_Differences_Among_Species_9-Aug-1999.

FAO, Health hazards associated with animal feed. FAO Rome, Section 1, http://www.fao.org/docrep/012/i1379e/i1379e01.pdf.

Fruncillo, M.D., Richard, J. (2011). *2,000 Toxicology Board Review Questions*, Xlibris Publishers, USA.

Gupta Ramesh, C. (2018). *Veterinary Toxicology: Basic and Clinical Principles*, 3rd ed. Academic Press/Elsevier, San Diego, CA.

Gupta, P.K. (2014). *Essential Concepts in Toxicology*, BSP Pvt Ltd, Hyderabad, India.

Gupta, P.K. (2016). *Fundamental in Toxicology: Essential Concepts and Applications in Toxicology*, 1st ed. Elsevier/BSP, London, UK.

Gupta, P.K. (2018). *Illustrative Toxicology*, 1st ed. Elsevier, San Diego, CA.

Gupta, P.K. (2019). *Concepts and Applications in Veterinary Toxicology: An Interactive Guide*, 1st ed. Routledge, London, UK.

Gupta, P.K. Ed. (2010). *Modern Toxicology: Adverse Effects of Xenobiotics*, Vol. 2, 2nd reprint. PharmaMed Press, Hyderabad, India.

Gupta, P.K. Ed. (2010). *Modern Toxicology: Basis of Organ and Reproduction Toxicity*, Vol. 1, 2nd reprint. PharmaMed Press, Hyderabad, India.

Gupta, P.K. Ed. (2010). *Modern Toxicology: Clinical Toxicology*, Vol. 3, 2nd reprint. PharmaMed Press, Hyderabad, India.

Klaassen, C.D. (2019). *Casarett and Doull's Toxicology: The Basic Science of Poisons*, 9th ed. McGraw-Hill, New York.

Klaassen, C.D., Watkins, III JB. Ed. (2015). *Casarett & Doull's Essentials of Toxicology*, 3rd ed. McGraw-Hill, New York.

Renwick, A.G. (2008). Toxicokinetic. In: Hays, A.W. (Ed.), *Principles and Methods of Toxicology*, 5 ed. Taylor & Francis Groups, Boca Raton, FL.

Sutmoller, P. (1997). Contaminated food of animal origin: Hazards and risk management. *OIE Scientific and Technical Review*. 16(2). http://siteresources.worldbank.org/INTARD/843432-1111149860300/20434404/ContaminatedFood.pdf.

Timbrell, J.A. (1997). *Study Toxicology Through Questions*, 1st ed. Routledge, London, UK.

INDEX

A

B

C

F

G

H

M

N

O

P

 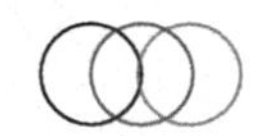

Q

R